Techniques in Pedology

To our wives

TECHNIQUES IN PEDOLOGY

a handbook for environmental and resource studies

R.T. Smith B.Sc., Ph.D.
and
K. Atkinson B.A., M.Sc., Ph.D.

Department of Geography, University of Leeds

ELEK SCIENCE
LONDON

First Published in Great Britain in 1975 by
Paul Elek (Scientific Books) Ltd.,
54-58 Caledonian Road,
London N1 9RN

ISBN 0 236 30939 0 (cased edition)
0 236 31020 8 (student edition)

Printed in Great Britain by
Unwin Bros. Ltd., The Gresham Press, Old Woking, Surrey
A member of the Staples Printing Group

CONTENTS

PREFACE

The techniques used in soil science are to be found dispersed throughout a considerable literature which presents a problem of information search for those studying and researching into soils. This book attempts to ease the problem by covering a wide spectrum of the techniques currently used in soil investigations, and seeks to strike a balance between activities carried out in the field and those normally pursued indoors, involving mapwork or laboratory analysis of soil samples. We hope the presentation of such a breadth of material within a single volume will be of value to instructors and students alike.

Soil science has only appeared infrequently in the past as a separate undergraduate scheme of studies in universities and advanced colleges. It has traditionally been a component of agricultural courses but is now increasingly appearing in the fields of botany, ecology, engineering, earth sciences and in physical geography. Although techniques cannot be thought of as having pre-determined applications, we have attempted to illustrate the kinds of practical studies which are feasible. For this reason, this selection of methods should certainly be of use to any undergraduate student of soil science and to students studying the soil as part of any other degree or diploma. Also, a selection of the methods will be found suitable for use in courses in education colleges and in senior or high school curricula. As pedologists and geographers, the authors have attempted to meet the requirements of the increasing numbers of students studying and researching into environmental problems, in addition to those traditionally of agronomy and engineering. This responds to the growing concern to understand the functioning of soils, ecosystems and all types of transfer process in the landscape.

The three sections of this book represent no more than convenient divisions, and it will be appreciated that they form a continuum of activities. Hence fieldwork is a precursor of all laboratory work at some stage, even though the study and interpretation of published soil maps need not actually be linked with any study on the ground. Soil maps may often seem to be produced as ends in themselves, but they may also be used as the rationalised starting point for a number of pedological inquiries, thereby generating, in turn, a fresh cycle of laboratory work.

In Part A the instructions on fieldwork are designed to help students to describe soil profiles in the field and to collect and prepare samples for analysis and indoor display. In addition, field skills include the ability to distinguish and map, on the basis of genesis and morphology, different soil types which may occur in close proximity. These soil units may be re-interpreted on the basis of their limitations for agricultural or other uses and set against a predetermined scale of capability. These various activities reflect the different areas of investigation current within studies of soil in the field.

Part B presents material which is not generally available in other texts. It discusses the value of soil maps and aerial photographs as sources of information and gives examples of exercises which in principle can be applied to any area for which there is adequate background information. For many students the soil map permits an appreciation of the distribution of soil types in a wider selection of areas than he can hope to visit, and thus allows him to appreciate associations of soils developed under different environmental conditions. Finally, in addition to providing clues to soil characteristics, aerial photographs are being increasingly used to facilitate and support field investigations.

Part C details a number of fundamental procedures in soil science without which the study of soils would be a largely subjective operation. These procedures include comments on theory as well as details of experimental procedure. Particular attention has been paid to the calculation stage of each method as the authors consider this to be a deficiency of existing sources, making methods initially less accessible to the beginner. It is worth pointing out that most of the techniques discussed can equally well be applied to the analysis of water, plant material, rocks and sediments. One recurrent problem in soil analysis is the availability of facilities and apparatus, and for this reason alternative methods are presented for those experiments which require more sophisticated instrumentation. Appendices B-F provide additional information to accompany Chapters 6 and 7.

The authors wish to thank all people who, by suggestions and discussion, have helped in the preparation of this book. In particular they thank the following for permission to use photographs: Meridian Airmaps Ltd. for Plates 4 and 5; Soil Conservation Service, United States Department of Agriculture for Plate 6, and Aerofilms Ltd. for Plate 3. We also thank Dr Brian E. Davies, Department of Geography, University College of Wales, Aberystwyth for permission to redraw Figures 7.3 and 7.7 from his originals. Our thanks are due to Dr E.A. Fitzpatrick for permission to reproduce Plates 7A, 7B and 8A.

PART A

FIELD INTERPRETATION AND SAMPLING

1 SOIL DESCRIPTION AND SURVEYING

1.1 SOIL PROFILE DEVELOPMENT

G. W. Robinson wrote that a distinction should be drawn between soil as a material and soils in their field context. It is the latter, concerned with the natural occurrence, classification and development of soils which has formed the central theme of pedology, a science owing much to the pioneering work of V.V. Dokuchaiev almost a century ago. Since the interest in soils from the outset was generated by the need to assess the extent of particular soil resources, pedology should not be regarded merely as the study of soils for their own sake. Hence other terms such as edaphology, introduced by Buckman and Brady (1969), referring to soils and plant growth, serve only to indicate the varying points of view which can be taken within the subject.

Soil in its natural state results from the weathering of minerals by physical, chemical and biochemical processes. It represents the theatre of activity between atmospheric processes on the one hand and the imperceptibily slow process of landform development on the other. Soil is therefore geologically ephemeral and is only ever preserved into a future age by exceptional circumstances of burial. Representing the medium of plant growth and to some extent the biologically degraded portion of the earth's crust, soil is coterminous with the lower boundary of the biosphere.

While soils vary greatly in their organic matter content, it is this connection with organic life which, above all, distinguishes soil from any other weathered residue or superficial deposit. A plant cover is, in any case, a prerequisite for soil stability. Soil material, as a consequence of its weathering history and biology, contains clay and humus, which enables it to resist rapid changes in moisture level. While constituting the nutrient reservoir of soils, clay and humus also help soil to retain essential elements against the influence of percolating water. These properties together with innumerable microbiological transformations enable the dynamic qualities of soil to be appreciated.

The vertical section through soil as it occurs in the field is known, perhaps paradoxically, as the *soil profile*. The changes in colour and consistency from top to bottom are normally recognisable as a sequence of layers or *horizons*. These horizons are characteristic of different soil types but it is important to realise that such diagnostic horizons may also take very long periods to develop. Distinctness of horizonation normally depends upon the balance between soil mixing mechanisms, such as provided by soil fauna, and the normally vertical movements of water and solutions.

Major factors contributing to the development of soil profiles are climate, vegetation and organisms, topography and the lithology of the *parent material* from which the soil is forming. All soil processes, which include those of weathering and redistribution of the products of weathering, result from the combined operation of these factors. The factors themselves, together with man's intervention are closely interrelated and in Figure 1.1 they have been organised into a system. In this diagram the central process domain, the soil matrix, is not really susceptible to division, but organic and inorganic aspects have been separated so as to make explicit the links with the external environment (Smith, 1973). Climate ultimately determines the types and rates of weathering and redistribution processes. It has a strong influence on available flora and fauna and is in turn modified by altitude and siting, and by vegetation

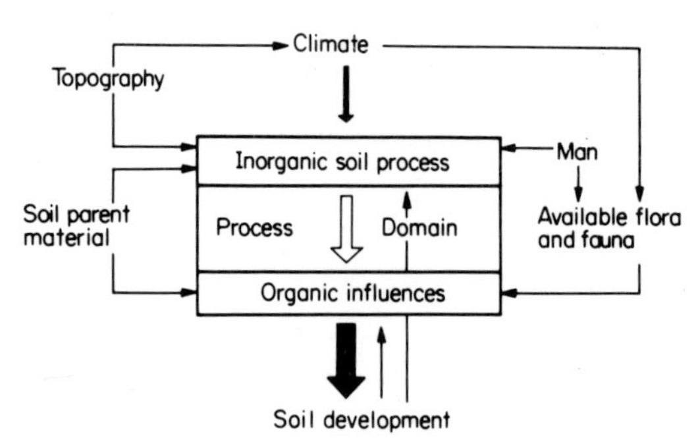

FIGURE 1.1 *The soil development system - some fundamental process and response linkages*

in terms of microclimate. But whether a particular flora and fauna colonises a site depends also on the chemical nature of the parent material. The latter, with siting, will determine the texture and drainage which are both crucial to the rates and types of soil process. The human agency is complex in its influence and ultimate effects, but man's principal modifications are in relation to the plant cover, to drainage and addition of chemicals. A characteristic of man's actions has been to accelerate processes. Hence increased crop growth resulting from the above practices has been beneficial to many soils, while soil erosion has often resulted from over-exploitation of mineral or moisture reserves in the soil, or the vegetation cover on its surface.

Soils are understood to evolve principally by a deepening of their profiles and by increasing differentiation of their horizons. For example, a brown forest soil with a largely homogeneous profile may evolve in time into a 'sol lessivé' with a heavier-textured B horizon as a result of acidification and consequent leaching of clay. Clearly, those properties which a soil acquires at any stage in its development must influence subsequent processes and hence the course of its own evolution. The thinner arrows on the diagram indicate such feedback or response mechanisms. For example, the gradual loss of bases and increasing acidity of a soil may initiate vegetation changes. Such events are believed to have occurred in the evolution of present-day heaths and moors but can be demonstrated in a contemporary setting on pastureland which is heavily grazed for a number of years without adequate treatment.

By considering soils on different scales it is possible to elucidate the controls which each factor exerts. On a world scale the major contrasts are climate-dependent. For instance leaching, or the removal principally of the soluble products of weathering, predominates in humid region soils of both high and low latitudes, but temperature determines the style of chemical alteration associated with it. Podzolic soils with a leached, *eluvial* A horizon and an enriched, *illuvial* B horizon are characteristic of higher latitudes and altitudes where, in addition to the common bases, sesquioxides (Fe_2O_3, Al_2O_3) are more mobile than silica (SiO_2). On the other hand, sesquioxide-rich, clay soils are widespread in equatorial and tropical regions where silica is more soluble. Many soils between these broad zones experience limited leaching or even the converse. Hence the soils of the great Steppes and Prairies commonly have B horizons of accumulation which contain carbonates and other more readily-soluble salts. On their drier margins and in arid regions, particularly where ground water is near the surface, saline and alkali soils occur. Capillarity, or the rise of moisture towards the soil surface, is a process which takes place spasmodically in all but waterlogged soils. In the absence of effective leaching it is also able to produce its own stratification of salts within soil profiles.

Soils in a particular locality will tend to vary with landform and hydrological changes, with vegetation cover or land use differences and, of course, with parent material. Most frequently it is the case that within an area of similar parent material, topographic variations provide the framework for soil detail. Although the precise relationship between topography and other factors depends mainly on climate, this notion, that of the soil *catena*, is of value in soil studies. It permits the grouping of neighbouring soils by association instead of classifying them on the basis of morphology alone. It also helps one appreciate that soils are a continuum across the landscape and only exceptionally offer striking boundaries.

There is another aspect to the question of soil assemblages being related to landform. Position on slopes affects the liability of sites to erosion or deposition and hence the probability of profiles having experienced burial or truncation in the past. Some soils may, for example, seem immature, while others display surface features which one would not associate with modern pedogenesis. Soil maturity implies the development of all features associated with a particular soil type or set of processes and is obviously, therefore, a highly subjective statement about any soil. It is clear that because of age differences in soils between low and high latitudes as a result of the Pleistocene glaciations, it would be unwise to assume that soils characteristic of major zones are a permanent expression of those environments. Some soils, chernozems for instance, have developed predominantly in fresh parent materials after the last glaciation while others, such as red Mediterranean earths and terra rossas, have developed under a succession of cool-moist and warm-dry climates since the Tertiary period.

The mineral assemblage too, whilst partly reflecting the present environment of weathering is also dependent on the ageing of the soil minerals. The likelihood of inheritance of soil profiles from former climatic conditions is suggested by the presence of incongruous horizons. Examples of the latter are the presence of an illuviated clay horizon in the soil of a presently dry area or the existence of a fragipan (a partially cemented and indurated layer believed to originate under periglacial conditions) within the parent material or C horizon of the soil. Alternatively, adjacent soils may exhibit very different degrees of weathering, such as those of the eastern United States where parts of the coastal belt have intensely weathered profiles, while glacial loess deposition towards the interior has given rise to altogether younger landscapes.

Before starting soil studies in the field one should be aware of the processes which have brought soils into being. Although, in a general way, the present environmental relationships with soils provide an adequate framework, one must anticipate that soils may have been subject to degrees of truncation or burial during the course of their long development. This possibility is just as important in a small area as on a world scale. It is a matter of common experience that soils selected at random are rarely textbook examples of particular soil types.

Soil survey comprises the following operations which the fieldworker carries out almost simultaneously.

1. *Description* - identification and description of soils within the mapping area
2. *Classification* - grouping of described soils into classes
3. *Mapping* - plotting the class boundaries on a base map.

Soil mapping has latterly played a significant part in our understanding of how soil properties relate to the earth's physical features and of how those properties affect agriculture. No matter what the scale of a soil survey, the object of producing the map is generally to enable predictions to be made about the rational use of land. Usually this has entailed the application of pedology to the requirements of a stable and sustained agricultural system, although soil maps are increasingly being used for other purposes, such as in forestry, engineering and watershed management.

Traditionally, soil survey has depended upon observations of soil properties in the field, with laboratory analyses merely providing confirmation of field observations analogous to prognosis and diagnosis in the field of medicine. However, the increasingly detailed definition of diagnostic horizons in national surveys (U.S.D.A., 1975; Avery, 1973) and in other systems (FitzPatrick, 1971) means that in many cases the mapping units or classes are themselves dependent upon laboratory analysis. Although routine analyses of profiles have been carried out for many years as a basis for agricultural advice, the above refinements are quite distinct. Obviously, they set limits to the element of personal judgement, which although a problem between workers also happens to be one of the chief attractions and challenges of fieldwork. At the educational and training level, however, it is best not to become too engrossed in the detail of more sophisticated systems but to select categories of soil which are suitable in relation to the size of area being studied.

1.2 PROFILE DESCRIPTION

In addition to covering the full range of readily identifiable features, soil descriptions should be comprehensible to workers in other parts of the world. For this reason efforts have been made to standardise the terminology and the procedures which are involved in arriving at decisions.

The following paragraphs list the various attributes of soils which collectively define individual horizons and profiles.

It greatly facilitates systematic description to use a standard form such as the one designed for Table 1.1. For survey organisations a form may be considered essential while for the student it plays some part in helping to establish a conscientious approach. It can of course be criticised, like many tables, for fostering a compartmentalised attitude, but it does ensure that a property is never accidentally overlooked (e.g. New Zealand D.S.I.R., 1962). In any event, a field note book should always be carried.

Profile No. .... Date .... Locality ....

Map Ref. .... Surveyor(s) ....

Elevation .... Topographic context ....

....

Vegetation/land use ....

Additional Notes ....

....

Profile Sketch	Depth	Horizon	Colour	Texture	Structure	Consistence	Drainage	Stoniness	Organic matter, roots and fauna	Nature of boundary	Field test dat		
											pH		

TABLE 1.1 *Field description form*

The equipment needed for soil description and survey can be listed as follows: a spade, a trowel or soil knife, a pick for stony soils, an auger of screw or corer type, a metal tape measure, a colour booklet (supplied by Munsell Color Co. Inc., Baltimore, Md. U.S.A. or by the I.S.S.S. (Appendix A)), polythene bags and labels for sample collection (see Section 2.1), a clinometer or level for slope measurements, a compass, a washbottle and hand lens for textural assessment, a field pH kit of colorimetric or electrical type and such reagents (Table 1.7 and Chapter 7) as may be considered helpful for diagnosis in the particular study area.

After the soil face has been prepared (see Section 2.1), a photograph, preferably in colour, is taken, but if this is not possible a pencil sketch should be made. This can often help to focus attention on combinations of features and facilitate the systematic description.

Soil *colour* is described by reference to Munsell notations. In the Munsell soil colour booklets each page represents a *hue* or the relationship of a particular pigment to the primary spectral colours. On each page appears a matrix of coloured chips, each of which is given a verbal colour description on the page opposite. The vertical coordinate represents lightness or *value*, while the horizontal coordinate represents the intensity or *chroma* of the pigment. Soil samples are compared with these colours until a nearest match is obtained. Thus a soil with a notation 10YR 5/6 has hue 10YR, value 5 and chroma 6, and is designated 'yellowish brown'.

Soil *texture* or particle size composition is assessed in the field with the fingers, and reference should initially be made to a set of guidelines such as those presented in Table 1.2 (see also, Table 1.4). Using the table, one should work from left to right, accepting the fact that the field condition may be moist initially. When wetting the soil from a washbottle, the sample should not be allowed to become saturated, but just sufficiently moist to allow the fingertips to detect coarser particles and permit deformation. Lumps of parent material e.g. limestone, or nodules, hinder progress and should, if possible, be eliminated beforehand. The silt grade can give rise to confusion. Silt grains are visible only under magnification and can also be felt on the roof or the mouth, whereas clay particles cannot. For soils designated as sand or sandy loam an attempt should be made to estimate the grade of the sand as coarse, medium or fine. For definition of grain size classes, reference should be made to Section 6.2. Some terms may need explanation, for example gritty refers to the rough feel of the coarser sand fraction. Friable implies that the soil is well aggregated but crushes without difficulty. The meaning of other terms will be evident after consideration of structure and consistence.

As with all procedures which are dependent upon personal judgement

Appearance with help from hand lens	Feel	
	Dry	Moist
Sand grains rare or absent	Very hard crumbs or prisms which crush with difficulty	Very sticky and plastic
Moderate number of sand grains evident	Hard, but smooth when crushed	Plastic but only moderately sticky
	More friable and gritty	Sticky and gritty
Some sand but silt grains predominating	Hard to friable - smooth when crushed	Smooth, silky or soapy - plastic without being sticky
Similar proportions of sand, silt and clay	Friable	Slightly plastic
Sand grains predominate	Soft and gritty	Slight cohesion
	Very gritty - almost entirely sand	Forms flowing mass
Rock fragments	Coarsely gritty and without cohesion	

TABLE 1.2 *Soil texturing guide*

Type	*Class* (mm)		
	Fine	Medium	Coarse
Platy	1-2	2-5	5-10
Prismatic - columnar if tops rounded	1-20	20-50	50-100
Blocky - angular and subangular	1-10	10-20	20-50
Granular	1-2	2-5	5-10
Crumb	1-2	2-5	5-10
Massive	1-2	2-5	5-10
Structureless or single grain	not characterised		
Grade	Weak - indistinct, barely recognisable in the field Moderate - well formed and evident in the field Strong - well formed and strikingly evident in the field		

TABLE 1.3 *Characterisation of soil structure*

Shaping		Textural term
Forms long threads which easily bend into rings	Takes clear finger prints and will polish	Clay Silty clay
	Sand grains rub off surface	Sandy clay
Threads will just form a ring		Clay loam Silty clay loam
Threads will not bend to form a ring		Sandy clay loam
		Silt Silt loam
Makes cohesive ball but forms threads with difficulty and which fracture when bent		Loam
Just able to form cohesive ball but cannot form threads when rolled		Sandy loam
Cannot maintain any shape		Sand
		Gravel

Dry or slightly moist soil		Moist or wet soil	
Crushing	Cementation	Stickiness	Plasticity
1. Loose 2. Soft 3. Friable 4. Hard 5. Very hard	1. Weak 2. Strong 3. Indurated	1. Non-sticky 2. Slightly sticky 3. Sticky 4. Very sticky	1. Non-plastic 2. Slightly plastic 3. Plastic 4. Very plastic

TABLE 1.4 *Soil consistence terminology*

there is really no substitute for experience and although use of a table provides a basis, even a few attempts usually show up imperfections in its design when studying the soils of a particular locality.

Soil *structure* refers to the arrangement of individual particles into larger aggregates or *peds* under field conditions. Structures in soil arise principally from the swelling and shrinkage behaviour of the clay and humus fraction, otherwise known as the colloids. Structures are absent in soils which are kept waterlogged for long periods and thus lack aeration. Plant roots also play an important part in developing and maintaining structural units, the size and shape of which are very dependent on texture. The shape terms are mostly self-explanatory. Platy development is usually a horizontal feature while prismatic or columnar structures are, as the names imply, elongated in a vertical sense. These structures, together with blocky and massive ones, are bounded by more or less flat surfaces while the smaller granular and crumb units are irregular. Soils characterised by these small, irregular units allow much more rapid infiltration of water and better root development than the essentially constraining characteristics of other structures. Massive structure consists of very large prisms or blocks and is usually found in moist, clay-rich soils, just as single-grain structure is a feature peculiar to sandy loams and sands.

When reporting the soil structure (Table 1.3) it is necessary to distinguish the type (shape), class (size), and grade (distinctness) of aggregates (Clarke, 1971). Structural forms can be mainly associated with either surface horizons or subsoil. Hence it is more common to find prismatic, platy, course blocky and massive forms in the subsoil while surface horizons, often due to ploughing, display a generally more broken and granular appearance.

Consistence is a term which refers to the cohesion of soil particles or aggregates in either the dry, moist or wet states. Handling consistency is, of course, fundamental to the finger assessment of soil texture, though some separate record should be made of the cohesiveness of the soil in whatever moisture state it occurs on exposure. No wetting or drying need therefore be carried out. The terminology has varied somewhat between workers. Table 1.4 presents a list of terms in common use (U.S.D.A., 1951; Soil Survey of Great Britain, 1960). The extreme left hand column in the table refers to crushing between thumb and finger, (1) having no resistance and (5) being uncrushable by this means. Soils may also exhibit cementation while indurated layers only yield to a blow from a hammer. Cementation by calcium carbonate, iron oxides and other substances is usually very obvious when present. Soils which are moist or wet are judged in terms of *stickiness* or the degree to which their particles adhere to the fingers, and *plasticity*, or their response to continuous deformation.

Drainage of soil profiles is described by the following range of terms. *Excessively drained* soils are the freely draining soils on knolls or 'shedding' sites which are distinctly drought-prone. *Freely drained* profiles are well-aerated and are either without mottling or have sparse mottling below 24 inches depth. *Imperfectly drained* profiles have mottling, especially along root channels, but do not have a gleyed horizon above 24 inches depth. *Poorly drained* profiles possess the grey colours of continuous gleying within the top 24 inches, while in *very poorly drained* soils gleying occurs throughout the entire profile and a peaty humus layer is often present on the surface. Drainage has an important influence on the structure, organic matter status, earthworm and faunal population, colour, growth potential and other properties of soils. It may often be influenced by parent material, yet is frequently controlled by topographic factors. It is therefore important to determine whether a site is moisture-*shedding*, *normal* or moisture-*receiving*. Although this is frequently assessed on the basis of experience or even on intuition, it is always best to examine a number of adjacent soil profiles before making a decision.

The drainage terms introduced earlier refer to whole profiles, with the implication that siting factors are principally involved. It is important to appreciate that there are many examples of soil profiles which do not conform to this idéa of gradation. Hence gleying of a surface horizon may be induced by marked textural and structural differences between topsoil and subsoil. In such cases, water is either impeded in its passage through the surface horizons or a water table becomes perched above a heavy or indurated subsoil. In addition, the terms *surface-water gley* and *ground-water gley* which appear in the literature are not easy to apply in practice. Strictly, if a profile receives the bulk of its water from the surface, and its texture or some profile feature such as an iron or clay pan impedes the through movement of water, it may be termed a surface-water gley. In practice, however, soils with gleying are often complex and one or other source of saturation may be overriding, depending on the season or on the immediate past weather.

The *stoniness* of soils is another important property which modifies the effects of texture, constituting either an improvement or limitation to the agricultural prospects of a soil. Table 1.5, based on the British Soil Survey Handbook (1960), distinguishes the characteristics which can be recorded.

Quantity		Size		Shape
Stoneless	0	Gravel	2 mm-1 cm	angular
Slightly stony	< 5%	Small stones	1-5 cm	subangular
Stony	5-25%	Medium stones	5-10 cm	rounded
Very stony	25-27%	Large stones	10-20 cm	shaley
Rock dominant	> 75%	Boulders	> 20 cm	tabular

TABLE 1.5 *Description of stoniness*

Organic matter status, faunal activity and roots should also be recorded. Organic surface horizons are given designations as in Table 1.6 and it must be made clear whether the organic matter is intimately mixed with mineral soil or occupies a distinct layer. If surface organic matter has a peaty consistency, and comprises distinct layers which result from the decay of different types of plants or which reflect varying humification, the following scale attributed to Von Post, may be used.

Slight humification

H1 Completely non-humified and free of dy (brown or yellow-brown particulate material that can be squeezed out with the water) - yielding only colourless water when squeezed.

H2 More or less unhumified and free of dy. Yields only yellow-brown water when squeezed.

H3 Very slightly humified with a small amount of dy. Yields muddy water when squeezed, but the peat substance itself does not pass through the fingers.

H4 Slightly humified peat and weakly dy-charged. Yields very muddy water when squeezed. Residue slightly plastic.

H5 Humified peat with a considerable amount of dy. Plant structure quite evident. Yields very muddy water when squeezed and some of the peat substance escapes through the fingers. Residue quite plastic.

Medium humification

H6-H7 Well humified peat and strongly dy-charged. Visible plant structures insignificant. Up to two-thirds of the mass passes through the fingers on squeezing. Residue consists chiefly of root fibres and wood etc., and is strongly plastic.

Strong humification

H8-H10 Very strongly humified and almost completely dy-charged. No vegetable structures visible under field conditions; more than two-thirds of the mass squeezes through the fingers.

The quantity of roots present in any soil horizon should be described as follows: absent, rare, few, common or abundant, and a subjective estimate made along these lines with a note being added regarding their nature and size.

Horizon characteristics		Symbol	
Fresh, undecomposed organic litter	seldom saturated	L	A_{00},O_1
Partially decomposed or fermentation layer	seldom saturated	F	A_0,O_2
Well decomposed or humified layer	seldom saturated	H	A_0,O_2
Peaty surface layer, usually wet		O	
Surface mineral horizon with organic admixture		A	A_1
Surface mineral horizon with strong humus admixture		Ah	A_1
Surface horizon modified by ploughing		Ap	A_1
Surface horizon of immature soils rooted by vegetation		(A)	
Eluvial horizon depleted of clay and/or sesquioxides		Ae	E,A_2
Transitional to B but more like A		AB	A_3
Transitional to A but more like B		BA	B_1
Subsoil horizon usually containing illuviated material		B	B_1 to B_3
Horizon containing illuviated clay (e.g. sols lessivés)		Bt	B_1 to B_3
Horizon with humus deposition (e.g. podzols)		Bh	B_1 to B_3
Horizon enriched with sesquioxides (e.g. podzols)		Bs	B_1 to B_3
Iron pan or layer of maximum iron depostion (e.g. podzols)		Bf	B_1 to B_3
Weathered subsoil horizon without evident enrichment		Bw,(B)	B_1 to B_3
Transitional to C but more like B		BC	B_3
Weathering horizon with resemblance to the soil above but not necessarily identical to its former parent material		C	
C horizon with fragipan		Cx	
Bedrock or substratum which may or may not consitute the type of material from which the soil originally formed and which may principally influence soil drainage		R	D

TABLE 1.6 *Soil horizon nomenclature*

Horizon boundaries either merge (change over a depth >5 cm) are narrow (2.5 cm) or are sharply defined (<2 cm). On the basis of this and other property descriptions the various horizons are labelled. Table 1.6 lists a number of horizons which are encountered particularly in humid temperate regions. In recent years horizon symbols have become internationally recognised, Table 1.6 representing a concensus of recent sources (U.S.D.A., 1975; Ball, 1960; Avery, 1964, 1973; Mackney and Burnham, 1966; Bridges, 1970; Canada Dep. Agric., 1970· Clarke, 1971). The alpha-numerical system dates back to 1951 and earlier (U.S.D.A.), while letter suffixes were used in Kubiëna's (1953) system for European soils. An important feature of any system based on symbols is that it must cater for non-committal labelling as well as for accurate horizon description. The lower-case letter suffixes in the table are generally those which are associated with one particular horizon. The following suffixes are added to any affected horizon: g - gleying, ca - calcium carbonate, cs - calcium sulphate (gypsum), sa - more soluble (mainly chloride) salts. Further suffixes are also in use in various systems, some of which duplicate the horizons already provided for in the table (e.g. Bir, Bfe for Bf). Suffixes may also be compounded if necessary (e.g. Aeg, Apsa, Bhs). Plates 1A and 1B illustrate the use of this system for two contrasted soil types.

Designation of horizons by such letter combinations will involve an element of interpretation since many symbols have an explicit or clearly implicit evolutionary connotation. If one can regard the procedure of description as even moderately objective, then the allocation of horizon names may seem to be, and often is, rather subjective and circumstantial in the absence of laboratory data. For this reason the field routine ought always to be supplemented by field tests unless formal laboratory work is relied upon as a backup, as already explained. Table 1.7 will provide a framework for the field testing which will also include the measurement of pH. A number of colorimetric kits are commercially available for the rapid testing of soil nutrients in addition to the tests suggested in the table. The tests in Table 1.7 are quick qualitative tests designed simply to indicate presence or absence. The chart could easily be extended to cater for special requirements. As an instance, the sodium azide test has been introduced for the testing of sulphides in the reclamation of colliery spoil (Kohnke, 1953). 3 g sodium azide (NaN_3) is dissolved in 100 ml 0.1N iodine solution. A small piece of soil is placed in a test tube and a few drops of azide solution added. Effervescence indicates production of nitrogen and the presence of sulphide. The numbers in the table refer to sections in Chapter 7 from which further details can be found. For measurement of pH in the field, reference should be made to Section 7.2.

Appearance of material	10% HCl	20% H_2O_2	Rhodizonate	10% NaOH	Separate NaOH / NaF	Silver nitrate	Separate KCNS / $K_3Fe(CN)_6$	Sodium azide	Other comments or criteria	Substance
White	Eff. (Section 7.3)									Carbonate
			Decolourised (Section 7.4)							Sulphate
						Precip-itate (Section 7.5)			rapidly soluble in H_2O, Taste	Chloride NaCl
									Insoluble	Silica
Black or dark brown		Eff.								Mangani-ferous
		Almost no eff.							Crushes and streaks	Carbon (charcoal)
Usually dark and wet	Eff.	Eff.						Eff.	H_2S smell	Sulphide
Light brown to dark brown	Addition of both causes eff.			Dark brown Na-humate solution	NaOH extract strongest				Organic-see humi-fication scale	Peat
				Variable brown solution	Tonal compari-son (Section 7.14)				Mineral soil - see stru-cture table	Indicates humus type
Ochreous, grey or mottled							Colour compari-son (Section 7.9)			Iron II and III

TABLE 1.7 *Field diagnosis chart* N.B. Eff = effervescence

PLATE 1 A. *Soil profiles excavated in the field. (A) A brown calcareous soil developed over limestone. (B) A peaty gley soil formed in glacial drift.*

1.3 FIELD MAPPING

The technique adopted in soil mapping depends on the scale of the eventual soil map. At scales of 1:5000 to 1:25,000 *detailed* mapping is realistic while at smaller scales *reconnaissance* mapping is more appropriate. Map scale is usually determined by the purpose of the work and should be just large enough to allow a clear depiction of soil data considered to be significant for a given purpose. In addition, as the complexity of the soil pattern or the intensity of present or potential land use increases, so will the scale have to be increased. For the student, a scale of 1:10,000 can be recommended as a suitable starting point. The detail of observations needed at larger scales and the generalisation required at smaller scales are both achieved best after initial experience has been gained.

The base map should show as much physical and cultural detail as possible so that the surveyor does not waste time fixing his position. Acquisition of appropriate airphoto cover is therefore recommended. Its additional cost is a small price to pay for the ease of mapping it gives. Field mapping should be preceded by gathering information on the solid and superficial geology of the area, and also the climate, land use and morphological features. A general appreciation of landscape history can also give enormous insight into the interpretation of soil patterns.

Profile pits are located to cover all *major* variations in parent material, slope, vegetation and land use; these are described as in Section 1.2 and sampled as in Section 2.1. The final written report on the soils will normally contain a detailed description of each mapping unit together with analytical data based on horizon samples. The sampling can be conducted during initial inspection of profiles or left until mapping has been completed, thus allowing more representative sites to be selected.

In addition to major profile pits, detailed survey requires many *inspection* profiles where the soil is either augered or a pit about one foot in diameter dug. The position of all profiles is marked on the base map and a brief description of inspection points is made, which will include details of depth, colour, texture, and stoniness. Where other distinguishing features such as free calcium carbonate or mottling are important, these will also be noted. Abbreviations should be used so that profile information can be recorded directly onto the base map. For example, a particular soil might be recorded as follows:

10 inches, grey, sandy loam, stony	10 Gy SL St
12 inches, brown, loam, stony	12 Bn L St
8+ inches, dark grey, loam, stony	8+ DGy L St

In the process of classifying the soils of the survey area, it is helpful to distinguish between the *taxonomic unit* and the *mapping unit*. The former is determined by the classification system being used while the latter reflects the scale and purpose of the mapping exercise. Most soil classification schemes are hierarchical in structure, with the lower order categories being the most relevant to field mapping. The taxonomic units used most widely in North America and Europe are the *great soil group*, the *soil series*, the *soil type* and the *soil phase*.

Great soil groups comprise soils with broadly similar pedological processes operating in them e.g. podzol or chernozem. The *soil series* comprises profiles with similar type and arrangement of horizons, developed from a particular parent material. Each series usually occupies a particular topographic position and it therefore forms the central concept and mapping unit for most detailed surveys. The *soil type* is a subdivision of soil series, based on the texture of the surface soil. Thus a component *type* of the Miami *series* would be the Miami silt loam, signifying a silty loam surface texture. The soil *phase*, on the other hand, is a subdivision of series *or* type on the basis of any characteristic considered significant to management. Thus Miami silt loam, stony phase and Greenwell, eroded phase are subdivisions of type and series respectively. Detailed surveys are always based on the mapping of individual series with their component types and phases. All these units are identified and mapped in the field. If any small area possesses a regular repetition, complicated pattern or mosaic of different soils it is customary to mark the area as a *soil complex*. This is an example of a mapping unit which comprises two or more taxonomic units. A further mapping unit is the *soil association*, which comprises two or more fairly repetitive soil series, one of which is usually dominant. The different soil series are usually related to topography. As a mapping unit, however, it is used in reconnaissance survey where scale and detail do not allow component taxonomic groups (e.g. series) to be distinguished. The soils found within an association do not necessarily belong to a single great soil group, and, like the complex, the unit is essentially one of mapping convenience. In Canada and Scotland, however, it is treated as a taxonomic as well as a mapping unit; here, the fact that component soils are linked by variations in topography and drainage is regarded as significant from a genetical standpoint.

Once the mapping units have been decided upon, the soils can be allocated to them as fieldwork proceeds. As an example of classification in action, it is instructive to consider the case of mapping soil series in humid temperate regions. Firstly, an assessment of soil parent material must be made at each profile site. A suitable symbol is then written on the map e.g. CL for Carboniferous limestone or TC for a glacial till derived from Chalk. This is an important first step as each soil series should have the same parent material. Additionally, in Canada and Scotland each parent material grouping is, by definition, a soil association grouping.

The next stage is to assess the drainage status of the profile. In many countries the drainage class subdivides the soils of one parent material into component series. For example, if three drainage classes are recognised these will delimit an equal number of soil series. Drainage class should be assessed as in Section 1.2, where five drainage classes are defined. Under such a system each parent material can have up to five component series, although such wide variation is rare and should certainly not be sought unless there is substantial topographic variation.

Soil boundaries can be drawn on the base map just as soon as the mapping units are established. It is clearly unwise to begin doing this early on before variations over a wider area have been studied. In detailed surveys the boundaries are always plotted in the field, whereas in reconnaissance surveys they are plotted later by extrapolation in the drawing office. The accuracy of boundary delimitation is, in fact, the main point of difference between reconnaissance and detailed surveys.

The quality of the finished soil map depends upon the accuracy with which the boundaries have been determined and plotted. Usually one soil category grades into another and only rarely does one find sharp boundaries. By convention this problem is recognised by permitting a mapped soil series to include up to 15% of any other (usually adjacent) soil series. To establish the location of boundaries, as many field traverses as possible should be arranged to cross, at right angles, major changes of relief, geology and vegetation. Many soil boundaries will coincide with observable features such as vegetation indicative of wet conditions, which helps the task of plotting, but such correlations should be continually tested, and in the final analysis all boundaries should be verified by digging or augering. It is not necessary to walk along all boundaries, but they should in the case of detailed surveys be plotted from observations made in the field. In reconnaissance mapping, the boundaries are more usually plotted on the field traverse lines, while between the traverses, boundaries are sketched in using indirect evidence provided by aerial photography or geomorphological facets. Reconnaissance mapping is therefore often more demanding of experience and skill than the detailed mapping based on straightforward observation and description.

Traditionally, soil survey has been conducted on a traverse basis which is *free* in so far as the areas chosen for more detailed consideration are a matter of personal judgement. A useful exercise is to organise simultaneously a survey of the same area based on a *grid*. Beckett and Burrough (1971) claim that the 'highly structured' survey is at least as accurate for flat or gently undulating agricultural terrain as conventional methods, and less time-consuming. It is also possible to see that the detail of fieldwork can more easily be adjusted to the scale of the resulting map, this being especially pertinent whenever photographic reduction is involved.

For a student exercise, the available time normally restricts the size of area which it is possible to study, but as much variation should be incorporated in as small an area as possible so that the principles can be illustrated quickly. A suggested grid could comprise 30-metre squares marked by tall posts. Students working individually or in groups can then approach the mapping of the prescribed area on either the free or grid sampling basis. After allowing adequate time for the fieldwork to proceed, a comparison both of progress and results can be made.

REFERENCES

Avery, B.W. *The soils and land use of the district around Aylesbury and Hemel Hempstead*, Memoir of Soil Survey of Great Britain, H.M.S.O. Harpenden, 1964.

Avery, B.W. 'Soil classification in the Soil Survey of England and Wales', *J. Soil Sci.*, 24(1973), 324-38.

Ball, D.F. *The soils and land use of the district around Rhyl and Denbigh*, Memoir of Soil Survey of Great Britain, H.M.S.O. Harpenden, 1960.

Beckett, P.H.T. and Burrough, P.A. 'The relation between cost and utility in soil survey', *J. Soil Sci.*, 22(1971), 466-89.

Bridges, E.M. *World soils*, C.U.P., London, 1970.

Buckman, H.O. and Brady, N.C. *The nature and properties of soils*, (7th Edn.) Collier-Macmillan, London, 1969.

Canada Department of Agriculture, *A system of soil classification for Canada*, Ottawa, 1970.

Clarke, G.R. *The study of the soil in the field*, (5th Edn.), Clarendon Press, Oxford, 1971.

FitzPatrick, E.A. *Pedology*, Oliver and Boyd, Edinburgh, 1971.

Kohnke, H. 'The reclamation of coal mine spoils', *Adv. in Agronomy*, 8 (1953), 317-49.

Kubliena, W.L. *Soils of Europe*, Murby, London, 1953.

Mackney, D. and Burnham, C.P. *The soils of the Church Stretton district of Shropshire*, Memoir of Soil Survey of Great Britain, H.M.S.O. Harpenden, 1966.

New Zealand Department of Scientific and Industrial Research, *Soil Survey manual*, Soil Bureau, 1962.

Soil Survey of Great Britain, Soil Survey Staff, *Field handbook*, 1960.

Smith, R.T. *Soil, environment and man*, Working Paper No. 5. University of Leeds, Department of Geography, 1973.

Tyurin, I.V., Gerasimov I.P., Ivanova E.N. and Nosin V.A., Eds., *Soil survey*, Israel program, 1965.

United States Department of Agriculture, Soil Survey Staff, *Soil survey manual*, Agricultural Handbook No. 18, Washington D.C., 1951.

United States Department of Agriculture, Soil Survey Staff, *Soil taxonomy, a basic system*, Agriculture Handbook No. 436, Washington D.C., 1975.

2 SOIL SAMPLING AND PREPARATION FOR ANALYSIS AND DISPLAY

Soil sampling should never be visualised as a separate operation but one which is linked to the purpose of an investigation, and which, in reality, forms part of a continuum of operations, commencing with the collection of soil and concluding with a sound interpretation of laboratory results. Many results are ruined by poor sampling through laziness or ignorance, and in this context it is well to recall the axiom which states that 'the analysis can be no better than the sample'.

2.1 SAMPLING FROM PROFILES

In mineral soils a pit is first excavated to the base of the profile. This can, however, be a difficult level to determine and in deep superficial parent materials one may have to define a functional base in terms of the limit of structure development or of biological processes in the broadest terms. Frequently this is an academic matter and some arbitrary limit, for instance 36 inches, may be set for the sampling sequence. The nature of the investigation usually suggests a realistic limit to sampling and indicates the need to sample particular horizons.

The profile face must be properly prepared, which, in the case of fine-textured soils, involves the detachment of small pieces with a knife or trowel to reveal natural structures and horizon boundaries. In the case of coarse-textured and friable materials, horizontal strokes of a builder's trowel usually prove most effective.

A representative soil specimen must be collected from each horizon. Individual projects will determine just how fine a definition is placed on the term 'horizon', and for this reason, to introduce any artificial features by the downward slicing of a spade, or to permit any contamination of lower horizons by topsoil, will be inimical to this objective.

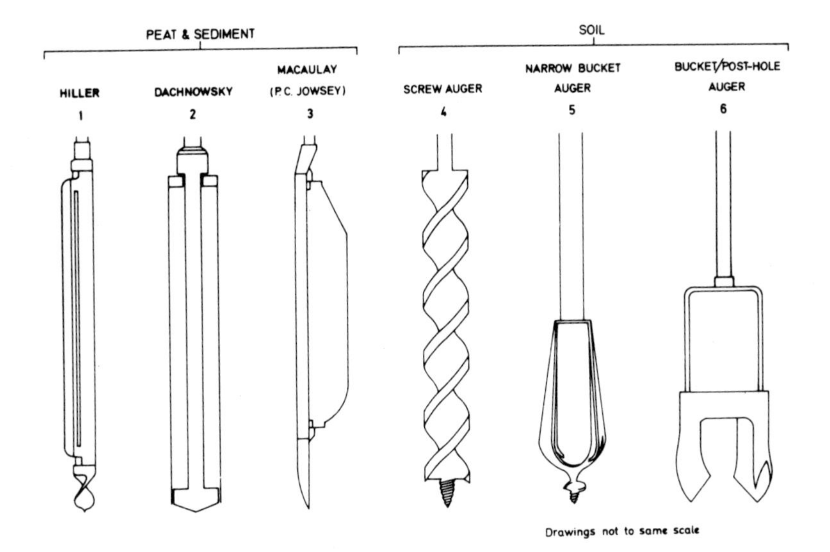

FIGURE 2.1 *A selection of sampling tools*

Although digging pits is the best form of insurance against the exceptional circumstance, when areal variations in horizon properties are being studied, various types of incremental sampler should be considered (see Figure 2.1). The most familiar is the screw auger which, although simple to operate, extracts only a small volume, the sample being deformed on the screw threads. Tubular corers provide an alternative, yet stoniness and plasticity frequently present them with insuperable problems. Very loose sandy material, whether wet or dry, is again virtually impossible to sample from depth unless one uses a chamber-type sampler. The post-hole auger is operational in a wide variety of textures, and because it brings up comparatively large samples from each depth it is often a realistic alternative to digging. An important point to bear in mind is that for any given depth, an adequate amount of soil must be collected for all the analyses envisaged. The larger the sample the more representative will it be, and the less significant will be errors of contamination which may arise through insertion of the sampler and by handling the sample.

It is absolutely vital that samples are labelled conscientiously at the time of collection. Individual systems may be considered foolproof but numbered tickets enable rapid identification in the laboratory. Cloakroom tickets are sufficiently durable to survive a limited period of moist conditions, the tear-off portion being placed inside the sample bag while brief descriptive information is inserted on the stub. The sample number can also be recorded against horizon descriptions in the field note book.

However, in cases where large numbers of samples are regularly processed it is advisable to give each sample a laboratory number which will not recur subsequently (Hesse, 1971).

When sampling peat and sediment the terrain will often preclude excavation, and unless one can find sections already exposed, use must be made of various kinds of chambered sampler (Figure 2.1). These allow cores to be extracted which are cylindrical or hemicylindrical depending upon the design. Depth measurements should be taken with care especially since peaty terrain sinks under the pressures applied to it in the sampling operation. Duckboards are easy to make and will prevent this. Many samplers, whether of a screw-down or push-down variety, have a nose section which causes disturbance to the layers beneath the sampled core. In such cases it is necessary to use two sampling holes - about 18 inches apart - so that each successive depth is sampled from alternate boreholes. There is no ideal sampler for all materials and because any instrument may meet difficulties at some depth in a sampling program, this is not a job to undertake single-handed. The cores are transported to the laboratory in plastic guttering or tubing and samples are then taken at any required interval (Faegri and Iversen, 1965; Wright *et al.*, 1965; Jowsey, 1966).

2.2 SAMPLING FROM AREAS

In this account, areal sampling refers to topsoil only. The kinds of problem envisaged are those of sampling from different soil types or areas of varying size which are under different land use or management (Jackson, 1958; Leo, 1963; Gallaher and Herlihy, 1963; Townsend, 1973).

Samples from cultivated land are taken from the top 6-9 inches, corresponding to the normal depth of ploughing. On grassland the 0-3 inch rooting layer provides an appropriate surface sample, while if pasture is to be brought under cultivation, a further sample to about 9 inches may be taken. In the case of arable land the sampling should be done between the harvesting of one crop and the sowing of the next, and for reasons of soil chemistry the samples are best taken before ploughing has occurred. Even under pasture it is advisable for standardisation if only because of seasonal fluctuations in soil properties. Also, to minimise the chances of naive interpretations, it is essential to obtain information (preferably quantitative) on the past as well as present treatments which have been applied to the area being investigated.

The best methods of taking samples employ either a spade, a 4-inch diameter post-hole auger or a 1-2-inch diameter screw auger. Although suggestions for exact sizes of samples appear in the literature and include the use of special frames to hammer into the soil, the best advice is to carry out the sampling for an individual project with the same tool (U.S.D.A. Soil Survey Staff, 1951). In any case, of far greater importance is the extent to which a single sample is representative of an area (or volume) of soil around it.

The ideal starting point in any sampling program is to identify as accurately as possible those areas which have reasonably uniform soils. These subjectively-defined areas are indicated initially by topographic variations, changes in soil tonality or differences in crop response, and an auger survey is essential. The areas of uniformity are known as the *sampling units* and an attempt should be made to keep their size as small as possible, ten acres being just manageable. Individual samples should be taken from randomly distributed points within these units and then mixed in the laboratory before the analysis. This *bulked* or *composite sample* should comprise equal amounts of soil from each of the sample points and it should weigh at least 1 kg. The greater the number of points, the more nearly will the bulked sample be representative, but as a general rule, for a uniform ten-acre area, a minimum of 25 auger or core samples should be taken.

Thus, a single spadeful of soil from a field should not be taken as representative of that area, even if the field is small enough to possess homogeneous soil conditions. Furthermore, if individual four-acre fields were ever regarded as satisfactory sampling units, the rapid increase of field sizes in recent years must surely encourage a more flexible framework within which sampling is to be carried out. In the past, inadequate fertiliser and manurial recommendations have certainly resulted as much from bulking samples from fields where soils were heterogeneous as from incorrect assumptions about the proportion of elements available to plants.

For suggested sampling procedures reference may be made to Section 5.6. The shape of the sampling units will determine the location of sample points. A regular area will suggest a grid framework, while an irregular or linear tract is best sampled at intervals along a series of zig-zag traverses (Jackson, 1958). Certain parts of fields are traditionally suspect and should be avoided; these include headlands where trucks have been driven and fertiliser bags dumped, as well as areas around trees, gates, ditches etc. But even with precautions such as these, it is probable that sampling systems which were adequate for relatively untreated fields will be insufficiently intensive for those which have been heavily treated with fertiliser, applied, for instance, in bands across a field (Melsted, 1967).

A recurrent problem of sampling is that one frequently has to work within the framework imposed by management units (farms and fields), so that a sampling design which obviates prior identification of physically uniform areas, yet which can be reasonably sensitive to local fertility variations, is of potential value. This is especially true of very large units which have little topographic or lithological variation. Such an opportunity is offered by a grid sampling method, and at the University of Illinois (Peck and Melsted, 1967) attention has been paid to the problem of sampling forty-acre fields which incorporate different soil types as well as variations in fertility. A scheme has been devised whereby each field is represented by 11 bulk samples, each of which comprises five separate, clustered core samples (Figure 2.2). Analytical results thus refer to a grid of 11 units rather than to areas of physical homogeneity. While more intensive sampling of the same area showed up imperfections in this design, it was clear that fertility patterns did not necessarily relate to soil type. This is a situation which must be anticipated on fertilised land, and which nowadays increasingly justifies the imposition of sampling units on the landscape rather than seeking guidance from the landscape itself.

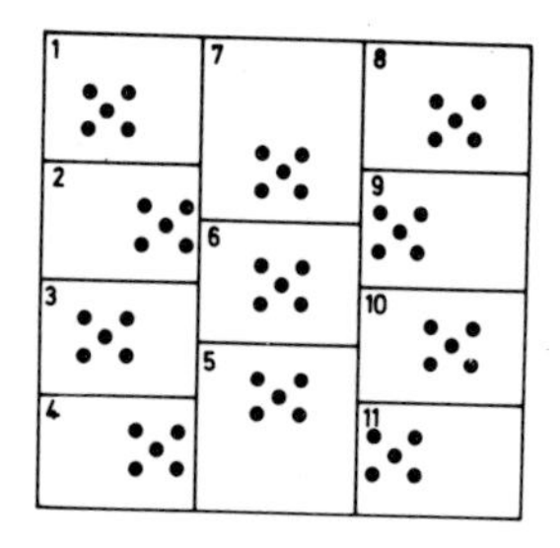

FIGURE 2.2 *Sampling design (after Peck and Melsted, 1967)*

2.3 PRETREATMENT FOR ANALYSIS

After collection, soil samples should normally be dried so that subsequent analyses can be carried out on a known weight of *fine earth*, comprising particles and aggregates smaller than 2 mm diameter. Generally, samples are *air-dried*, under which circumstance they merely reach a moisture equilibrium with the laboratory atmosphere. Some determinations (e.g. loss on ignition) may require that figures be based on *oven-dry* soil, but since several tests may be carried out on a sample, air-drying is the usual precursor to storage. The drying can be expedited by breaking up large clods and spreading the sample out on a sheet of polythene or metal foil. The temperature should not be higher than 35°C (Hesse, 1971) as organic and mineral transformations may otherwise occur which adversely affect a number of determinations. For some analyses any drying or delay over analysis militates against accurate results, or indeed any result. Estimations of ferrous iron, nitrate-nitrogen and a number of biological studies fall into this category. Some tests for soil constitution (engineering tests) and studies of soil fabric in thin section may dictate that field samples are kept *intact* as opposed to being reduced to a *disturbed* state as for the majority of analyses (British Standards Institution, 1967). Material with a high silt or clay content is best not dried because of the problem of subsequent disaggregation. In such cases (providing the material is nearly all fine earth) the sample is weighed out in the moist state and a moisture determination carried out on a replicate sample. Samples or cores of peat are best stored moist and cool, or preferably frozen before analysis so that any hardening and decomposition is avoided.

Drying of soil samples is followed by *crushing*. An ordinary ceramic mortar and pestle will be satisfactory, but, ideally, an agate mortar should be used with a wooden or rubber-ended pestle. In the majority of cases, simple crushing will not abrade primary particles, but pulverising or grinding is to be discouraged. Crushing may even sometimes be carried out on the bench top with a rolling pin. Crushed soil is placed on a 2 mm (10 mesh) sieve. Soil is shaken or brushed through the apertures, but not rubbed through. Material failing to pass the meshes comprises *gravel* and *stones*, and providing this genuinely consists of discrete fragments and not crushable aggregates, it can be weighed and expressed as a percentage of the soil mass. The nature of this coarse material often gives an indication of the parentage of the soil. If, however, it comprises more than about 20% by weight, care should be exercised when interpreting analytical results confined to the fine earth fraction. The coarse material is rarely retained. The fine earth which has passed through the sieve is stored in labelled bags or preferably glass, screw-capped bottles to await analysis.

2.4 SUBSAMPLING FOR ANALYSIS

Material weighed out for analysis must be representative of the already processed field material and not simply what was uppermost in the storage bottle. The merits of a consistent sampling technique at this stage are easily and frequently overlooked. Particles of varying size will tend to settle into bands within a storage container, with coarser particles appearing to rest on finer material. The importance of obtaining a representative *subsample* lies not only in the fact that the fine earth may represent a bulked or composite sample from the field, but that the composition of the different-sized particles will not be the same.

The two main methods of subsampling are those of *quartering* and the use of a *riffle-box* or *sample-splitter* (British Standards Institution, 1967; Hesse, 1971). The latter is mainly used where large volumes of soil are involved and consists usually of a metal box with a row of alternate chutes which deliver soil equally into two removable trays. The sample is thus split into halves until a subsample of appropriate weight is obtained. It is often convenient to reduce the size of air-dried samples in this way before obtaining the fine earth, but the dried, granulated soil must have been thoroughly mixed beforehand.

Quartering is less sophisticated and involves spreading the fine earth onto a piece of paper, dividing the soil into four approximately equal parts and returning the soil of two opposite quarters to the storage container. The remainder is mixed and spread out again, the process being continued until the required weight of soil (e.g. 100 g) remains on the paper.

2.5 SOIL MONOLITHS

Soil profiles which have been transported to the laboratory are commonly known as monoliths. As such, they may have as little appeal away from their field context for pedologists as dead specimens have for many biologists; yet, for educational and exhibition purposes they can be highly successful. Their main value lies in illustrating variations in structure, colour and other special features not only in soils of the immediate locality but of other regions to which visits cannot be made.

There are numerous accounts of how to collect and preserve standard soil monoliths which recommend the use of frames or boxes for fitting round a protruding portion of the soil face (Kubiëna, 1953; Taylor, 1960; Beater, 1963; Clarke, 1971). There may often be difficulties in collecting profiles which are stony or unstructured. The common mistake is to try to collect too large a specimen (Orsenigo and Cline, 1955). Often, for comparative purposes, it is the topsoil and upper subsoil alone which are diagnostic. In the laboratory, the soil face is treated with a resin which prevents it from crumbling and preserves its natural colour, though some authors have regarded this as a job to be undertaken in the field (Smith and Moodie, 1947).

In the case of profiles which are very deep or difficult to collect, an alternative solution is to take specimens which represent as nearly as possible each horizon, and store these in a partitioned box. However, if a monolith is to be a correct reduction, each horizon must be scaled down in proportion. If partitioning is unacceptable, care must be taken to avoid introducing artificial boundaries at points where the various pieces have been joined. In this respect a soil lacking distinct horizonation is a bad candidate for treatment in this way. Hammond (1974) has recently described a method for impregnating and mounting peat monoliths.

Another approach to monolith collection is known as the lacquer-film or resin-peel method. The profile to be collected is sprayed with a penetrative resin and a base board or piece of thick hessian is then stuck to the prepared face before finally peeling or easing the treated soil away from the surrounding profile (Lyford, 1939; Van Rooyen, 1959; Jager and van der Voort, 1966). Despite certain claims (Brand, 1959) this method is only well suited to relatively stone-free soils and those which, if not dry at the time of excavation, are likely to dry quickly and allow the resin to permeate. The technique can, of course, be carried out indoors on a monolith collected in the conventional way. One very useful method, which yields rigid monoliths of 1-1.5 inches thickness, of manageable weight, and which are easy to store, has been described by Clarke (1962) and this might be considered as a good compromise between the two methods outlined.

Two further methods involve the use of fine soil only and are not monoliths in the strict sense. They enable the scale of profiles to be greatly reduced, and while not showing structure or precise field colours nevertheless give good contrasts over a range of soils. The first approach utilises large test tubes or ungraduated measuring cylinders (e.g. 1000 ml size). Samples must be taken contiguously down the profile (e.g. for each 3 inch increment), air-dried, crushed and if necessary the gravel or stones removed, though this will depend on the size of vessel being used. The sample is thoroughly mixed and reduced to a 1 inch or 1 cm height in the vessel.

In the second method, contiguous samples are again collected from each inch or greater interval, dried, crushed, and made to pass through a 1 mm sieve. This fine soil is then dusted onto adhesive PVC tape so that 1 cm on the tape represents each contiguous sample. The tape is then mounted on a neutral grey background. Accepting its textural bias, this can be a rewarding way of presenting soil profiles in reports and on wall displays and lends itself to photographic reproduction. Furthermore, the dried soil may be compared with an ignited sequence of the same samples, thus allowing the effects of certain pedogenic processes (e.g. podzolisation) to be clearly illustrated (Dimbleby, 1962).

In all these methods, the final presentation of the profile should be properly annotated from observations made in the field. This is particularly important with the latter two methods as it may be difficult to avoid interfaces occurring between adjacent layers.

2.6 SOIL SIMULATION

Simulation experiments permit the observation of processes (e.g. percolation, Section 6.6) in granulated or sieved soil, actual soil cores or ideal media, and provide opportunities to artificially induce or re-create pedological phenomena. Replication with different media or solutions will also allow the study of rates of processes. The attraction of this form of experimentation lies as much in its ability to make visible an otherwise 'invisible world' as in striving to accelerate those processes which in the landscape have taken considerable lengths of time to achieve their effects. Whether research or demonstration orientated, such experimentation effectively provides 'living monoliths' in which the observation of each stage is quite as important as that of the end result. The following is concerned exclusively with the use of soil columns to demonstrate particular soil processes.

Perhaps the simplest example of simulation arises in the determination of exchangeable cations with the use of a leaching column (Section 7.6). The latter device, with sophistication, has been used for studies on the relative potency of plant extracts and synthetic agents in mobilising soil constituents such as free iron oxides (Hallsworth and Crawford, 1965) and in the movement of fine particles (Wright and Fors, 1968). But, although the leaching process is important in pedogenesis, it is not demonstrated quickly in a visual way in most soil materials and therefore ideal media are more often used. The latter can be made uniform, graded or comprise alternating bands of coarse and fine particles.

Easier to demonstrate quickly is the effect of capillarity, drawing up a solution of salts towards a heated surface and causing deposition of materials at varying heights according to solubility (Krupkin, 1963; Felitsiant, 1966; Szabolcs and Lesztak, 1967). Equally, one may attempt to create such features as carbonate concretions by limited leaching experiments (Hallsworth and Crawford, 1965) or induce soil structural changes by leaching a well aggregated soil with dilute sodium hydroxide. The latter will produce phenomena resembling solonetzic soils.

In experiments designed to induce pedological horizons, optimum size of column is about 2-4 inches in diameter; wider than cation leaching or chromatography columns and between 1 ft and 3 ft in height. The question of edge-effects on the glass and the effect of a warm laboratory on the entire column serve to underline the impossibility of achieving perfect simulation. Larger tanks may provide an answer to some of these problems but the wider the vessel the greater the problem of even packing and of uniform application of water or solution to the surface (Graham-Bryce *et al.*, 1967; Mansell *et al.*, 1968). Three examples of simulation experi-

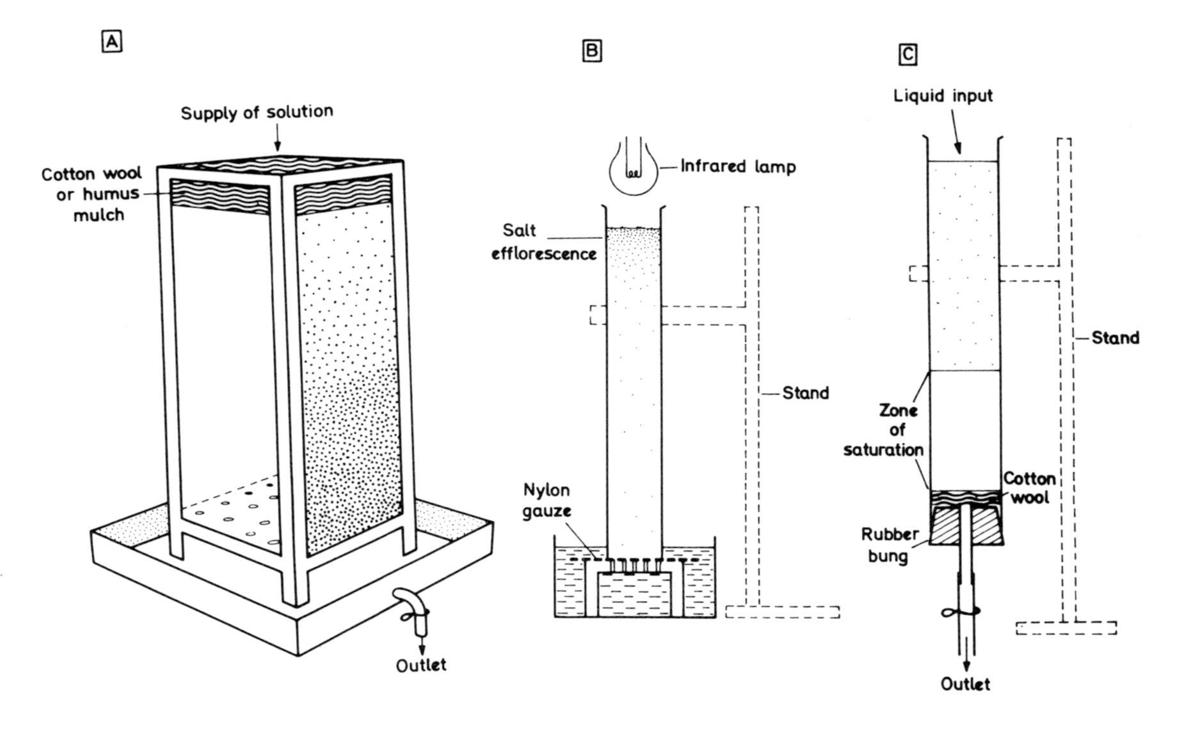

FIGURE 2.3 *Experimental design for simulation studies*

ments are illustrated in Figure 2.3.

In Figure 2.3, drawing A illustrates a device made of angle iron and plate glass with a basal grid having a nylon gauze overlay. This is a suitable set-up for inducing podzolisation of ferruginous sands given the application of a chelating agent. 0.02N EDTA or *Mimosa* (wattle) bark extract can be used as chelating agent; from 100-250 ml per day will generally be a sufficient volume of liquid for this size of apparatus. The volume of liquid applied is of greater importance to the success of this experiment than the concentration of reagent used. This is because firstly, simulation does not attempt to accelerate pedogenesis by any given amount but merely permits clearly visible changes to occur over a short period, and secondly, a purely practical consideration is that the features to be induced must appear within the vessel and not be leached out of it! It is therefore recommended that small amounts of reagent are applied at regular intervals, for instance on a daily basis, and that the column be kept fairly cool and out of sunlight. Simulation may be useful for inducing a number of pedological features but the influence of abnormal temperature profiles within the column and the possibility of photochemical reactions on the edges of the vessel must warn the experimenter against undue optimism.

Apparatus A (Figure 2.3) is also suitable for observing the effects of gleying on the underlying ferruginous material when a surface mulch is kept saturated. Furthermore, if the device is allowed to stand in a salt solution then, irrespective of the nature of the medium being used, the effects of capillarity will be observed within the column. For this purpose the mulch can be removed and heat applied to the surface as necessary.

Drawings B and C (Figure 2.3) are of thick glass cylinders secured by stands. As illustrated, B is set up for the capillarity demonstration, while C is a simple leaching column able to perform the same work as A. However, by closing the outlet tap at the base, saturated conditions and even those of a fluctuating water table can be induced in the lower part of the column to simulate the effect of gleying. Although the tank in A can be sat in a water bath, the set-up in C is more convenient for this particular demonstration.

It will be appreciated that control of water-flow into and through any simulation experiment is crucial to its success and this alone may, in the early stages, be a major preoccupation of the investigator.

In simulation, much is to be gained by having an experiment running in duplicate so that one can observe the results as experimental conditions are varied. Then, what might simply have been an attractive exhibit also becomes an educational and scientific experience.

REFERENCES

Beater, B.E. 'Soil profiles for display purposes', *J. Soil Sci.*, 14 (1963), 262-66.

Brand, F.C. 'The preparation of lacquer-film profiles of naturally moist soils' (in German), *Zeitschrift Pfl. Ernähr Düng*, 86 (1959), 123-31.

British Standards Institution, *Methods of testing soils for civil engineering purposes*, B.S.1377 (1967).

Clarke, G.R. 'The preparation and preservation of soil monoliths of thin section', *J. Soil Sci.*, 13 (1962), 18-21.

Clarke, G.R. *The study of soil in the field*, (5th Edn.), Clarendon Press, Oxford, 1971.

Dimbleby, G.W. *The development of British heathlands and their soils*, Oxford Forestry Memoir No. 23, Clarendon Press, Oxford, 1962.

Faegri, K. and Iversen, J. 'Field techniques' *Handbook of paleontological techniques*, Eds. B. Kummel and D. Raup, Freeman, San Francisco, 1965.

Felitsiant, I.N. *Regularity of capillary movement of water and salt solutions in stratified soils – an experimental study*, Israel Program, 1966.

Gallaher, P.A. and Herlihy, M. 'An evaluation of errors associated with soil testing', *Irish Journal of Agricultural Research*, 2 (1963), 149-67.

Graham-Bryce, I.J., Davies, R.I. and Al-Rawi, A.A.H. 'A simple and inexpensive apparatus for automatically controlling solution flow rates through soil columns', *J. Soil Sci.*, 18 (1967), 39-41.

Hammond, R.F. 'The preservation of peat monoliths for permanent display', *J. Soil Sci.*, 25 (1974), 63-6.

Hallsworth, E.G. and Crawford, D.V. *Experimental pedology*, Butterworth, London, 1965.

Hesse, P.R. *A textbook of soil chemical analysis*, John Murray, London, 1971.

Jackson, M.L. *Soil chemical analysis*, Constable, London, 1958.

Jager, A. and van der Voort, J. 'Collection and preservation of monoliths from sandy soils and ripened clay soils above and below the water table', *Soil Survey papers* No. 2, Netherlands soil survey Institute, Wageningen, 1966.

Jowsey, P.C. 'An improved peat sampler', *New Phytol.* 65 (1966), 245-48.

Krupkin, P.I. 'Movement of salt solutions in soils and soil materials', *Soviet Soil Science*, (1963), 567-74.

Kubiëna, W.L. *Soils of Europe*, Murby, London, 1953.

Leo, M.W.M. 'Heterogeneity of soil of agricultural land in relation to soil sampling', *J. Agric. and Food Chem.*, 11 (1963), 432-34.

Lyford, W.H. 'Preservation of soil profiles by Voigt's method', *Soil Sci. Soc. Am. Proc.*, 4 (1940), 355-57.

Mansell, R.S., Nielsen, D.R. and Kirkham, D. 'A method for simultaneous control of aeration and unsaturated water movement in laboratory soil columns', *Soil Sci.*, 106 (1968), 114-21.

Melsted, S.W. 'The philosophy of soil testing' in *Soil testing and plant analysis*, Soil Science Society of America, Special publication No. 2, 13-23, Madison, 1967.

Orsenigo, J.R. and Cline, M.G. 'An adjustable micro-monolith sampler', *Soil Sci. Soc. Am. Proc.*, 19 (1955), 99-100.

Peck, T.R. and Melsted, S.W. 'Field sampling for soil testing' in *Soil testing and plant analysis*, Soil Science Society of America, Special publication No. 2, 25-35, Madison, 1967.

Smith, H.W. and Moodie, C.D. 'Collection and preservation of soil profiles', *Soil Sci.*, 64 (1947), 61-69.

Szabolcs, I. and Lesztak, V. 'Capillary movement of sodium salt solutions in soil columns', *Soviet Soil Sci*, (1967), 483-88.

Taylor, J.A. 'Methods of soil study', *Geography*, 45 (1960), 52-67

Townsend, W.N. *An introduction to the scientific study of the soil*, (5th Edn.), Arnold, London, 1973.

United States Department of Agriculture, *Soil Survey manual*, Agricultural Handbook No. 18, Washington D.C., 1951.

Van Rooyen, C. 'An improved technique for preparing monoliths of sandy soils', *S. Afr. J. of Agric. Sci.*, 2 (1959), 256-66.

Wright, H.E., Cushing, E.J. and Livingstone, D.A. 'Coring devices for lake sediments' in *Handbook of paleontological techniques*, Eds. B. Kummel and D. Raup, Freeman, San Francisco, 1965.

Wright, W.R. and Fors, J.E. 'Movement of silt-sized particles in sand columns', *Soil Sci. Soc. Am. Proc.*, 32 (1968), 446-48.

3 JUDGEMENTS OF SOIL AND LAND QUALITY

3.1 INTRODUCTION

Most work in the fields of soil description and soil mapping is carried out for the practical purpose of improving man's management of the land. Traditionally, soil surveys have been regarded as necessary bases for agricultural development, but nowadays their range of application has extended widely to cover such uses of land as forestry, wildlife management, recreation planning and engineering construction. In the past, the soil scientist generally left the task of interpretation to specialists in the relevant field - the agricultural officer, the engineer or the biologist. The disadvantage of this lies in the fact that the soil scientist has studied the soils closely and is clearly the best person to make accurate predictions of their performance.

Increasingly, soil survey reports now include interpretative judgements on the soils they describe. Usually this includes not only an evaluation of soil profiles and soil properties for specified uses, but also judgements of land quality by *land evaluation*. In this context 'land' is defined as the totality of environmental factors operating at the earth's surface, be they climatic, geomorphological, pedological or biotic. Thus the soil profile is recognised as providing only partial criteria for judging land quality. Plant growth is also affected by climatic and surface characteristics that should be recorded together with soil information. The soil surveyor describes these as *site characteristics*, and Table 3.1 outlines the land parameters that should be recorded. This information is regarded as additional to that recorded in the profile description (Section

1.2) and is used in conjunction with it in making assessments of land quality.

Location	
Elevation	
Relief	slope gradient, length, shape, pattern
Drainage	surface drainage, water-table depth
Aspect	
Micro-relief	
Parent Material	
Climate	rainfall regime, growing season
Stoniness	surface, profile
Vegetation	
Erosion	
Rooting Depth	

TABLE 3.1 *Check list of site parameters*

Although the range of techniques used in land evaluation is increasing as more and more specialists appreciate the utility of soil data, three basic approaches can be distinguished in this general field. Firstly, the *landscape* approach enables a map of land *potential* or land *capability* to be constructed for any area. A generalised map is quickly produced (Section 3.2), although the most popular technique remains that of the Soil Conservation Service of the U.S.D.A. (Section 3.3). This is based on a detailed soil map and is suitable for scales as large as 1:10,000. The second approach to land evaluation, the *parametric* approach, produces numerical ratings for soils with respect to their suitability for specified uses. Two commonly used systems are those of Storie (Section 3.4) and Clarke (Section 3.5). The success of any system of land evaluation ultimately depends on the accuracy with which it can predict yield. Thus the third approach, by *yield assessment*, allows an experimental design to be established which will relate yield to soil and site characteristics (Section 3.6).

A further point needs to be emphasised concerning land and soil judgements. As the *actual yield* from any given area clearly depends on management, as well as factors of climate and soil, it is conventional to assume optimum management in making the judgements. Thus, *potential yield*, rather than *actual yield*, is being assessed. It is important to realise this principle on account of the influence it has on the choice of land and soil attributes that are weighed in the evaluation. It will be noticed that systems of land judgement rely heavily, but not exclusively, on physical properties of soil. For example, soil texture receives detailed consideration, whereas nutrient content is often ignored completely. Here, the assumption is that under optimum management, the land user would normally maintain chemical fertility by such operations as liming and the use of artificial fertilisers. In this sense, nutrient content would not be a limitation on yield. Limitations due to soil texture would not normally be ameliorated by management, and hence would collectively be considered as a limitation on potential yield.

Chemical factors only become important in land evaluation when the cost of remedying the limitation is likely to be high in relation to the economic yield of the land. The chemical limitations most commonly met are depression of yield by salinity, an excessive concentration of salts in the soil solution, and by alkalinity, an excess of the sodium ion on the soil exchange complex. An example of where these factors need to be considered is given in Section 3.4.

3.2 GENERALISED AGRICULTURAL ASSESSMENT OF LAND AND SOIL

As a preliminary to detailed work, a method of introducing the student to the grading of land on the basis of physical quality is to construct a generalised land classification map of a relatively large area, e.g. an entire 1:250,000, 1:63,360 or 1:50,000 topographical map. An initial exercise is best carried out for an area where soils information, preferably in map form, is available and where climatic data can be consulted. In addition to soil and climate, topography is the third criterion considered in the classification and hence a suitable topographical map is required. At the reconnaissance level of the classification, the closeness of map contours is a sufficiently accurate measure of slope.

The guide adopted for the exercise is that used by the Ministry of Agriculture, Fisheries and Food (1966). This classes land into 5 grades according to factors of climate, relief and soil. The grades recognised are:

Grade	I	Land with very minor or no limitations
Grade	II	Land with some minor limitations which exclude it from Grade I

Grade	III	Land with moderate limitations
Grade	IV	Land with severe limitations
Grade	V	Land with very severe limitations

A suggested procedure for producing a reconnaissance land classification map is to consider limitations in the order: slope, climate and soil. A piece of tracing paper is placed over the topographical map of the area to be classified. Areas where slope affects land quality are demarcated according to the following scale:

Degree of slope	*Highest possible Grade*
Less than 1 in 8	I
1 in 8 to 1 in 5	III
1 in 5 to 1 in 3	IV
More than 1 in 3	V

The climatic factor can next be considered. For England and Wales, rainfall, evapo-transpiration, temperature and exposure are taken into account to give the following guidelines:

Climatic situation	*Highest possible Grade*
(West) Land over 400'; more than 45" rainfall	III
(East) Land over 400'; more than 40" rainfall	III
Land over 600'; more than 50" rainfall	IV
Land over 1000'; more than 60" rainfall	V

The simplicity of the climatic factor is governed by the need to devise a system suitable for a national-scale survey and by the paucity of data on meso-climates. Areas where the climatic limitations are relevant are traced onto the classification map.

The final and most testing stage of the exercise lies in appending soil limitations to those of slope and climate. For this, it will first of all be necessary to study the units shown on the soil map, together with any written report. A convenient way to do this is to write down a list of the main soil classes in the area, and alongside, to add notes on physical limitations for each soil. The limitations will primarily be concerned with wetness, structure and texture, depth, and stoniness. These limitations can either restrict the range of crops, depress yields, cause variability in yield or increase cultivation costs. The Ministry of Agriculture, Fisheries and Food (1966) gives extensive examples of the characteristic parent materials, situations and soil series of each Grade. The following guidelines allow one to consider each limitation in turn:

a) *Wetness*

Imperfectly drained	Not I
Poorly drained	III or IV
Very poorly drained	V

b) *Structure and texture*

Loams, sandy loams	I
Silt loam to silty clay loam	II or III
Clay loam to clay	III or II
Sand to loamy sand	IV to II
Peat, peaty loam-drained	I to III
Peat-undrained	V

c) *Depth*

Less than 10"	Down one grade

d) *Stoniness*

Stones over 50% of soil	Down one grade
Large, unmovable boulders	IV or V

In addition to producing a finished map of agricultural land classification, it is advisable to summarise the rationale behind the system for the benefit of any potential user. This can be done by making notes in tabular form under the column headings: soil category, suggested Grade, reasons for Grade. If a land classification map of the area has been published, it would of course be instructive to compare the finished map with the published map.

3.3 THE PREPARATION OF MAPS OF LAND CAPABILITY

A land capability classification is an interpretative grouping of the soils of any particular area for land use (mainly agricultural) purposes. It judges the capability of land from physical factors of climate, site and soil on the one hand, and what is known of crop response to management on the other. The most widely used system is that developed by the Soil Conservation Service of the U.S.D.A. (Klingebiel and Montgomery, 1961) which has been modified and adapted to the needs of many countries, for example in Canada (Canada Land Inventory, 1965), in Europe (e.g. Bibby and Mackney, 1971) and in developing countries.

In most modifications of the U.S.D.A. system, land is grouped into 7 classes, which are arranged according to their suitability for agricultural use. Class 1 has no important limitations for a wide range of agricultural crops, while limitations increase from Class 1 to Class 7. Class 7 has most limitations and is unsuitable for agriculture. In general, Classes 1, 2 and 3 are considered suitable for the sustained production of crops, Class 4 is not suitable for sustained crop production, and Classes 5 and 6 are considered unsuitable for crops but suitable for permanent pasture.

The capability classification is based on certain assumptions which must be understood before attempting to interpret the climatic, site and soil data. These assumptions are:

1. Capability is assessed under good management practices and not necessarily under present management.
2. Each class may include different kinds of soils, often requiring different management and treatment. *Class* only indicates the severity of limitation.
3. Soils considered suitable for improvement (e.g. by drainage or some stone removal) are classified according to future limitations after improvement. Major land reclamation may change the capability classification, as may also any change in knowledge of fertility.
4. The system is entirely based on environmental criteria, and economic and social factors such as farm structure and cultural patterns are not considered. Of the environmental factors, physical limitations are rated more severely than chemical ones, since the former are more difficult to upgrade.

Following assumption 2, it will be clear that a division of classes is needed in order to provide information on the kind of limitation. This is achieved by dividing classes into *subclasses*, which form the main mapping unit. The specific limitation is indicated by a reference letter which is suffixed to the Class number, e.g. Subclass 3e indicates land of Class 3 where erosion damage (e) is a problem. Class 1 will have no subclasses.

The number of subclasses recognised, and their letter designations varies slightly from country to country. Table 3.2 shows three common systems, which can be further modified and added to, as required. In the description of capability subclass, it is possible to note as many limitations as occur. However, on a map it is conventional to use no more than two subclass symbols for any one class. It is also conventional to list symbols in order of importance and, if two limitations are equal in weight, to use the priority e, w, s, g, c.

A. *U.S.D.A.*

e	erosion
w	excess water
s	soil limitation
c	climatic limitation

B. *CANADA LAND INVENTORY*

d	soil structure and/or permeability
e	erosion
f	inherently low fertility
i	inundation
m	soil drought
p	stoniness
r	shallow depth
s	combination of soil limitations
t	topography
w	excess water, excluding inundation

C. *SOIL SURVEY OF GREAT BRITAIN*

w	wetness
s	soil limitation
e	erosion
c	climate
g	gradient and soil pattern (e.g. Complex)

TABLE 3.2 *Designation of subclass limitations*

The details of the criteria for designating capability subclass are listed in Klingebiel and Montgomery (1961) and Bibby and Mackney (1971), and the details relevant to the area being studied will need to be consulted and discussed before work starts. It is convenient, however, to have a readily accessible guideline to hand and Table 3.3 is presented for this purpose. It is adapted from Bibby and Mackney (1971), and indicates the highest Class a site can belong to if it has a certain limitation. The following examples will illustrate this point:

Class	Climate	Gradient	Drainage (Wetness)	Stones	Effective Rooting Depth (inches)	Erosion Hazard
1	F	$< 7^{\circ}$	Good	None	> 30	F
2	F	$< 7^{\circ}$	Good-imperfect	None-slight	> 20	F
3	F,U	$< 12^{\circ}$	Imperfect-poor	None-stony	> 10	F
4	F,U	$< 16^{\circ}$	Poor	None-stony	> 10	U
5	F,U,S	$< 25^{\circ}$	Poor-very poor	None-very stony	> 8	S
6	S,U,F	$0-25^{\circ}+$	Very poor	Extremely stony	< 8	S
7	S,U,F	$0-25^{\circ}+$	Very poor	Extremely stony	< 8	S

F - favourable; U - moderately unfavourable; S - severe

TABLE 3.3 *Guidelines for the designation of capability classes*

A soil with a depth of less than 8" can be no higher than Class 6

A soil which is stony can be no higher than Class 3

A site with a severe climate can be no higher than Class 5

These guidelines will clearly need modification in areas outside a humid maritime temperate climate. The climatic limitation in particular is governed by characteristics of growing degree-days or available moisture, or a combination of the two. Table 3.3 can also be adapted to serve as a blank data sheet for recording capability information in the field; this can best be done by deleting column 1 and adding additional columns for Site number, Soil class, Soil texture, Other soil limitations, Present land use, Capability class and Additional comments.

Plate 2 illustrates how two landscapes can be divided into capability subclasses. The landscape of part of the English Pennines shown in Plate 2A has potential only for pasture land. This has been subdivided on the basis of heavy texture (s), gradient (g), liability to erosion (e) and wetness (w), according to the scheme of the Soil Survey of Great Britain. Plate 2B shows the junction of grey wooded (light tones) and grey-brown podzolic soils (darker tones) in Saskatchewan, Canada. Using the designation of the Canada Land Inventory, it has been subdivided on the basis of structure limitations (d) and stoniness (p).

3.4 THE EVALUATION OF LAND ACCORDING TO THE STORIE INDEX

The index for rating soils which has been developed by Storie (1954) has become known as the Storie Multiplication System, and has been widely used in North America and elsewhere. The Storie Index is a numerical rating of the degree to which a particular soil has physical, chemical and biological properties suitable for the growth of crops; it is basically an index which reflects the soil's potential utilisation and productive capacity.

Three principal factors are considered in computing the Storie Index of a soil unit. These are: (a) the character of the soil profile as reflected in texture, structure and inherent fertility; (b) topography, and (c) other modifying factors such as climate, salinity/alkalinity, degree of stoniness and susceptibility to erosion. Each of these three factors (a, b and c) is rated on the basis of 100 for the most favourable or ideal conditions, with progressively larger subtractions from this as conditions become less favourable for plant growth.

In North America, the Storie Index of a particular soil unit is often given in the soil survey report and memoir. As the soils

(A)

(B)

PLATE 2 *Subdivisions of landscapes into capability subclasses.*
(A) Part of the English Pennines in Co. Durham.
(B) Part of the Canadian Prairies, Saskatchewan.

are mapped and classified, the character and degree of development of the soil profile, the soil reaction and colour, the physical composition of the various horizons, the composition of the parent material, the profile drainage and surface relief are all considered in detail. In deriving the Storie Index, soil surveyors use their experience and judgement derived from studying soils and agriculture in a particular area over a considerable time. Although it may be difficult to reproduce this experience in a class exercise or field project, it is very rewarding to attempt a quantitative rating, as this will underline the need for precision in studies of land potential and the necessity for avoiding rather loose qualitative judgements. Although index ratings do not eliminate subjectivity, they certainly help to keep it to a minimum level. The value of indices as *comparative* measures must again be emphasised.

The Storie Multiplication System is not a rigid system which is applied directly and unchanged to any area. Depending on the soil and climatic conditions of the region under study, it will be necessary to rearrange some factors and to add others in order that a more accurate expression of prevailing environmental conditions is reflected in the rating. The following system is presented as a guideline and has been developed by the Soil Survey of Saskatchewan, Canada (Ellis, Acton and Moss, 1968). The interest in this region is to assess spring wheat production under dry-farming conditions. Three main factors are employed (A, B and C), and each of these factors has a maximum rating of 100 points.

Factor A	*Soil profile*	*Points*
1.	Texture throughout	40
2.	Structure throughout	30
3.	Inherent fertility a) depth	15
	b) colour	15

Factor B	*Topography*	*Points*
1.	Flat to depressional	10-80
2.	Gently undulating	90-100
3.	Moderately undulating	80-90
4.	Very gently rolling	70-80
5.	Gently rolling	60-70
6.	Strongly undulating	40-60
7.	Moderately rolling	40-60
8.	Strongly rolling	30-40
9.	Hilly	20-30
10.	Steep slopes	10

Factor C	*Modifying characteristics*	*Points*
1.	Climatic favourability	25
2.	Salinity and alkalinity	25
3.	Stoniness	25
4.	Liability to wind erosion	25

If one is working without the benefit of long experience of soils and crop yields in the area under study, it will be clear that the use of the Storie System will provide a good test of the student's knowledge of agronomy, and is well worth attempting for this reason. The key parts of the exercise lie in the initial definition of the characteristics to be rated for each factor and the scale of subtraction from the maximum point allocation. Once these two points have been decided, use of the system will give confidence and increasing expertise.

The results of the work can either be expressed by ratings annotated on a soil or base map, or by constructing a table (Table 3.4). If information on crop yields is available either from official statistics or farmers, exercises on the degree of correlation between present and potential productivity are clearly possible.

Soil Association	Soil Type	Ratings Factor A	Rating Factor B	Ratings Factor C	Index A × B × C
Sceptre	Clay	35-25-30	100	15-25-25-10	90 100 75
	Clay loam	33-26-28	95	10-25-23-15	87 95 73
Fox Valley	Clay	30-30-25	100	10-25-25-15	85 100 75
	Loam	20-20-20	90	10-25-25-10	60 90 70
Haverhill	Clay loam	32-26-25	90	10-25-20-20	83 90 75
	Light loam	15-25-20	90	10-25-20- 5	60 90 60

TABLE 3.4 *Storie indices for selected soils*

3.5 RATING SOILS ACCORDING TO THE CLARKE INDEX

A rating index that is eminently suitable for evaluating soil quality in the field is the Clarke Index (Clarke, 1951, 1971). This points system gives a numerical rating to a soil profile on the basis of the physical characteristics of texture, depth and drainage status. A soil profile pit is dug to a minimum depth of 30 inches if possible, and descriptions made of profile characteristics (Section 1.2). Soil texture, depth and drainage status are determined and recorded in the usual way, and this information is used to compute the Clarke Index.

Clarke's Index is defined as: profile value = texture value (V) × drainage factor (G), where texture value (V) = depth (D) × texture rating (T). The texture rating (T) is derived from the field textural designation (Table 1.2) according to the following scale:

Texture	*Value*	*Texture*	*Value*
Sandy loam	16	Silt loam	15
Loam	20	Silty clay loam	12
Clay loam	18	Silt/silty clay	6
Gravel	3	Clay (structured)	15
Fine sand	8	Clay (massive)	5
Coarse sand	14		

It will be noticed that in the case of clay, allowance is given for its structural condition in addition to texture. Where clear textural horizons exist in the 30" profile, the texture value (V) is computed by measuring the depth of each textural horizon and multiplying each depth by the relevant texture rating, e.g.

15" clay loam over 15" massive clay

$$\text{Textural value } (V) = (15 \times 18) + (15 \times 5) = 270 + 75 = 345$$

Where soil depth is 30" and more, the textural rating is multiplied by 30, e.g.

30" of loam

$$\text{Textural value } (V) = 30 \times 20 = 600$$

Where the profile is less than 30", but the parent rock is shattered and weathered and is basic in character, e.g. limestone or basic igneous, the disintegrating horizon is counted as one third of the texture rating of the particles in the matrix. No compensation is given for solid rock.

Depth of Gleying (inches)	*Drainage Factor*
9–12	0·5
13–15	0·6
16–18	0·7
19–24	0·8
25–30	0·9
No gleying	1·0

The drainage factor (*G*) is calculated according to the presence or absence of gleying. The rating is based on a value of 1·0 for perfect drainage, with incremental subtractions according to the depth at which gleying occurs. The final profile value is the product of the textural value (V) and this drainage factor (G). For 30" of loam with gleying at 23", the profile value is

$$600 \times 0{\cdot}8 = 480$$

The attraction of the Clarke Index lies in the fact that only three fundamental physical properties are used, and that each of these can be determined relatively easily in the field. It is an index which lends itself particularly well to correlation and regression analysis in relation to figures for crop yields if these are available. Clarke (1951, 1971) has related index ratings to wheat yields and proposes three soil quality classes (First, Second, Third) based on index and yield data. There are interesting possibilities for adapting and devising a classification for any specific crop in any specific area. In addition to indicating relationships between good crop yields and good quality land, questions of climate and management are inevitably raised when there are large discrepancies between soil quality and yield. These can form the basis for further detailed studies on land utilisation.

It is also an instructive exercise to experiment with modifications in the soil properties used in Clark's original index. For example, such characteristics as stoniness and micro-relief might merit attention in a particular area. In using one's ingenuity to devise and test modifications one has to balance the need to construct as meaningful an index as possible, with the need to retain the principles of simplicity and field assessment that are embodied in the Clarke Index.

3.6 CROP YIELDS AS INDICES OF LAND AND SOIL PRODUCTIVITY

The success of any method of judging land and soil quality depends upon the accuracy with which it predicts soil productivity, i.e. crop yields, under specified land use and management conditions. Whilst yield data may be obtainable from sources such as agricultural censuses, farmers' records and advisory service experimental plots, the coverage and detail of such statistics is often very uneven. The necessity may arise for one's own data on yield to be collected. The assessment of soil productivity by field experimentation lies at the basis of agricultural advisory work and correspondingly, has a vast background literature. For the newcomer to the field, reference to Avery (1962), Townsend (1973) and Cooke (1967) is advised. The method described here has been used with success by Dermott, Roberts and Wilkinson (1965), and allows one to collect yield data from a number of different soil or land classes. The necessity for collecting results that are sufficiently detailed to permit statistical analysis is compromised slightly by the necessity for studying a group of sites at once. This is achieved by simultaneously studying a range of 'benchmark' sites.

The mapping units of interest are chosen and small plots set up at two sites on each. The number of soil units selected can vary between about four and eight, and each plot is approximately four yards square. Each plot is subdivided into four equal micro-plots which allows rotation over four years. A stock-proof barrier and wire netting around the experimental area will reduce animal and bird damage.

All experimental areas grow the same crops. The choice of these will clearly depend on the climate of the area being studied and previous experimental experience by advisory officers. A suitable choice for eastern England is potatoes, barley, Italian ryegrass and kale, a choice which incorporates two cash crops and two forage crops. The assessment is based on total dry matter production. All four crops are grown simultaneously each year, so that after four years all four crops have been fully rotated on each micro-plot.

The crops are grown under optimal management. This demands a high standard of cultivation using hand methods, and requires good seedbed preparation, weeding, control of pests and disease, and the elimination of any nutrient limitations by adding optimal levels of major and minor nutrients. Essentially, the yields will represent *potential* soil productivity rather than *actual* soil productivity, and measure the combined effects of climatic and soil physical parameters.

At harvest time the crops are hand-harvested, washed of any adhering soil particles, dried and weighed. After four years, yield data will give a mean dry matter yield for each experimental plot. The duplication of experimental plots will give a mean yield for each mapping unit. This mean figure is taken as a relative index of the potential productivity of the unit in question.

The problems of carrying out field experimentations are mainly those related to finding landowners willing to allow the use of their land in this way and to the high demands of care and management required by the experimental plots. However, the great advantage of having a *quantitative* measure of land capability is ample justification for the effort involved. The plots can be expanded to become major educational and research projects over a continuing period of time. Many refinements will suggest themselves and can become incorporated into the work. For example, regional climatic data can be supplemented by on-site meteorological instruments, e.g. a soil thermometer and rain gauge. Equally, the potential productivity of the soil types investigated can be compared to the yields recorded by local farmers on the same soil.

The index

$$\frac{\text{Actual yield per acre}}{\text{Potential yield per acre}} \times 100$$

thus becomes a comparative measure of soil management, and is similar to the average (A) and high (B) yield figures in many American soil survey reports that give estimated yields for specific crops on given soil types.

REFERENCES

Avery, B.W. 'Soil type and crop performance', *Soils and Fertilisers*, 25 (1962), 341-44.

Bibby, J.S. and Mackney, D. *Land use capability classification*, Technical Monograph No. 1, Soil Survey, Rothamsted, 1971.

Canada Land Inventory, *Soil capability classification for agriculture*, Ottawa, 1965.

Clarke, G.R. 'The evaluation of soils and the definition of quality classes from studies of the physical properties of the soil profile in the field', *J. Soil Sci.*, 2 (1951), 50-60.

Clarke, G.R. *The study of the soil in the field*, (5th Edn.), Clarendon Press, Oxford, 1971.

Cooke, G.W. *The control of soil fertility*, Crosby Lockwood, London, 1967.

Dermott, W., Roberts, E. and Wilkinson, B. 'The use of soil maps', *Nat.Agric. Advisory Service Quart. Rev.*, 69 (1965), 16-22.

Ellis, J.G., Acton, D.F. and Moss, H.C. *The soils of the Rosetown map area*, University of Saskatchewan, Saskatoon, 1970.

Klingebiel, A.A. and Montgomery, P.H. *Land capability classification*, Soil Conservation Service, Agricultural Handbook No. 210, Washington D.C., 1961.

Ministry of Agriculture, Fisheries and Food, *Agricultural land classification*, Agricultural Land Service Technical Report No. 11, 1966.

Storie, R.E. 'Land classification as used in California for the appraisal of land for taxation purposes', *Trans. 5th Int. Cong. Soil Sci.*, 3 (1954), 407-12.

Townsend, W.N. *An introduction to the scientific study of the soil*, (5th Edn.), Arnold, London, 1973.

PART B

CARTOGRAPHIC STUDIES OF SOIL PATTERNS

4 INVESTIGATIONS USING AERIAL PHOTOGRAPHS

4.1 THE AERIAL PHOTOGRAPH AS A TOOL IN SOIL STUDIES

Clarke (1971) has made a useful distinction between the use of aerial photographs for purposes of identification and their use for interpretation (API). In the broad field of soil science, the aerial photograph is used in both of these ways, in the sense that the investigator wishes both to identify all features of the landscape that are of value in his soil studies, be they physical or human, and to interpret soils data from the image of the photograph.

The aerial photograph is used by soil scientists and soil surveyors in three distinct, but not mutually exclusive, ways. Firstly, the photograph is often used as a base on which to plot soils information and to draw boundaries between soil classes, using information on the soils collected in the field. By showing the totality of physical and human landscape features, the aerial photograph usually allows soil mapping to proceed more quickly and more reliably than working with a topographic map alone. The topographic map may not contain sufficient information, for example in upland areas, for accurate soil boundaries to be drawn (Ball, Hornung and Mew, 1971).

Secondly, as the aerial photograph shows many features of the soil's genetic environment (particularly landforms, vegetation, drainage pattern and, to a lesser extent, geology), it gives important clues for interpretation of the origin and distribution of soils. For example, by giving a bird's-eye view of the soil cover in relation to landforms, observations on the relationships between the distribution of terrain and soil patterns can supplement those observed by ground survey. Indeed, at all map scales it has been repeatedly shown that the aerial photograph reveals patterns that would not normally be distinguished by fieldwork alone. An example of this in alluvial areas is given by Evans (1972).

The third category of the use of aerial photographs in soil studies concerns the interpretation of soil features from the photo image. Generally this is an indirect exercise in which inferences and deductions are made that relate the image either to single soil properties (e.g. drainage status or stoniness) or to defined taxonomic units (e.g. soil associations or soil series). In many ways, this is a more difficult use of aerial photographs, and inferences made usually depend on the investigator's knowledge of soil relationships in the area under study and on his general experience of soils and aerial photographs. In carrying out exercises of this type it is advisable to start initially with areas that are relatively well known and then to proceed to regions where one's knowledge is of a reconnaissance nature or entirely gleaned from soil maps and memoirs. The choice of local areas allows field checking to be carried out after the interpretation exercises have proceeded as far as they can indoors. In this way the advantages and limitations of aerial photographs can be quickly identified.

As the study of aerial photographs has become such an important part of soil and landform mapping, it is vital to recognise its limitations, particularly with regard to some extravagant claims that have been made regarding the value of the whole range of remote sensing techniques. It is important to realise that the aerial photograph shows soil surface features only. The *influence* of sub-surface features *may* be shown, but the photograph itself cannot directly show the soil's three-dimensional characteristics and particularly its variations with depth. It follows from this that the aerial photograph cannot be regarded as a substitute for the field study of soils, despite claims to the contrary. Clues to soil characteristics in any area depend to some extent on a knowledge of the soils in the field, whether directly or through primary sources. This emphasises the complementary and indirect roles that aerial photographs play in soils work.

A final principle that needs to be borne in mind concerns those landscape elements that the photograph shows relatively clearly, namely, relief, vegetation and land use. It is important that these elements, when used in correlations with soil attributes, are not merely used in a deterministic manner. Each is an independent variable in its own right, in Jenny's (1940) terminology, controlled by many influences. Interpolation of soil features from such visible elements can only be reliably made when the relationships of climate, landform, vegetation and soil are confidently known for the area under study. An example of the kind of misinterpretation that should be avoided is provided in some arid regions where vegetation patterns are clearly demarcated on aerial photographs but where such patterns may not be correlated with any identifiable soil characteristic.

4.2 KEY PROPERTIES OF AERIAL PHOTOGRAPHS

Vertical aerial photographs of contact size (mostly 9 × 9 inches) are the most generally used in soil studies. Single prints can be studied by eye, particularly where the terrain is very familiar to the interpreter, but for close analytical work it will be necessary to study stereoscopic pairs with the aid of a simple pocket stereoscope. This not only gives three-dimensional details but also gives a magnification of between two to three times. Simple instructions for achieving stereoscopic vision will be found in Clarke (1971) or Dickinson (1969). The normal overlap of prints is 60% and stereo-pairs can normally be located by making reference to the index of the flight line and the serial number of the print. If a topographic map is available at approximately the same scale as the aerial photograph, the extent of areal coverage can be determined quite easily.

In addition to flight and exposure identification numbers, the aerial photo contains information on the time and date of exposure. This can prove to be very valuable information in deciphering land use patterns, for example, by relating the photo images to stages of crop growth. For many aspects of soils work photography in spring and autumn is more useful, as much of the soil in arable areas will then not be covered by crops.

The scale of the aerial photo is determined by the flying height of the aircraft and the focal length of the camera lens. The relationship is:

$$\text{Representative fraction} = {}^{f}/_{h}$$

where f is the lens focal length and h the flight altitude in the same units. On older aerial photos this information is given, but on more modern prints it is now usual to give the representative fraction as well. When making scale measurements from the photograph, it is important to remember that all aerial photos have image displacement. This may be slight in relatively flat areas but becomes serious as the amplitude of relief increases. In hilly areas, it is negligible only at the centre of the photograph.

There is no single scale of photography that satisfies all the requirements of soil studies. Whilst different tasks ideally require different scales, in practice, photo coverage is generally so incomplete that, unless aerial photography is specifically commissioned, one will usually be working with scales of 1:10,000, 1:20,000 and 1:60,000.

Fortunately these are useful scales for many purposes. Scales as large as 1:5000 and 1:10,000 are excellent for studying detailed soil variations and micro-relief in small areas such as individual fields and farms. Scales smaller than 1:20,000 are usually very good for studying regional soil patterns, particularly with respect to their relations with drainage and physiography for example, but are clearly less useful for more detailed work. Table 4.1 shows how scale affects the usefulness of aerial photographs to the soil scientist.

	Scale		
	1: 2,500 to 1:10,000	1:10,000 to 1:20,000	Smaller than 1:20,000
Natural features			
Regional	Poor	Good	Excellent
Local	Excellent	Good/fair	Poor
Minute	Excellent	Poor	Very poor
Cultural features			
Regional	Poor	Good	Fair
Local	Excellent	Fair	Poor
Minute	Excellent	Poor	Very poor
Soils			
Regional	Poor	Excellent	Good
Local	Good	Good	Poor
Minute	Excellent	Good	Very poor

TABLE 4.1 *Depiction of landscape features according to scale* (adapted from American Society of Photogrammetry (1960))

The increasing range of applications of aerial photo interpretation over the last ten years has been paralleled by technical developments in photo production and publication. Whilst monochrome panchromatic film remains the mainstay of aerial photography, the use of other types of film is increasing in popularity and, with a decrease in cost, could expand dramatically over the next decade.

At present, four main types of film are available. These are:

a) *Panchromatic*, which produces a monochrome (black and white) photo
b) *Natural colour*, whose advantages may or may not offset the higher costs
c) *Infra red monochrome*, normally used with a red filter
d) *Infra red colour* (false colour) which is sensitive to the infra red regions of the spectrum and accentuates differences between different spectral bands

Useful background information on the properties and potentialities of the different types of aerial photo films is given in Tarkington and Sorem (1963), Mott (1966) and by Kodak Ltd., (1969). As the cheapest and most readily available prints remain panchromatic monochromes, these will form the basis for discussion in this chapter.

The techniques of aerial photography are developing rapidly and methods currently in the experimental stage of development could quickly become routine procedures.

Multispectral photography, whereby the use of filters allows a record to be made of different wavelengths and sensitivities, offers most promise for soil interpretations. Details of this and other advanced techniques are given in Hunter and Bird (1970).

For any given scale of photograph, six properties of photo images are useful for purposes of interpretation. These are: (a) tone or colour; (b) texture, i.e. the mosaic created by tonal variations over a small area; (c) pattern, i.e. trends or lineaments larger than those in (b); (d) shape; (e) size; (f) association of objects.

In monochrome prints, all natural colours appear black, white or shades of grey. Thus differences in soil properties - for example, colour, wetness, texture, depth, stoniness - will be reflected in differences in the *tone* of the image on the photograph, if they are to be observed at all. The tone of an image is determined by the amount of light reflected by the image; those objects that appear white are reflecting most of the light hitting them, whilst those that appear dark are reflecting only a small amount of light. Factors other than the roughness and colour of the reflecting surface can affect the amount of light reflected - for example, the angle of the sun's rays and the angle of reflection from the object to the camera. A classic instance of this is provided by lakes, which reflect most light in a single direction; they appear white if the light is reflected directly into the camera and black if the camera lens is outside the line of reflected light. Usually, however, general tonal relationships can be identified which aid interpretation. Bare, dry soil will often appear light, reflecting most incident light in many directions, whilst bare, moist soil will appear as a darker tone of grey. In fact the tonal variation between moist, dark areas and drier, light areas is one of the most widely used tools in the interpretation of soil patterns. Soil drainage variations, once detected, can usually be ascribed to topographic variation, e.g. poorly drained hollows and abandoned drainage courses, or parent material differences, such as impervious tills and freely-draining outwash sands. These factors will usually be detectable from the photo by stereoscopic examination of topography and physiographic pattern.

An additional property often discernible from the aerial photograph is soil texture on account of the influence which this soil property has on soil moisture relations. Providing other factors are not overriding, fine-textured soils are dark in tone whilst coarse-textured soils are light. Thus heavy clay, whether glacial, lacustrine or alluvial in origin, will appear dark grey, whilst contiguous areas of sands or sandy loams often appear white to light grey in tone. The areal pattern shown by differences in tone in such situations is clearly a useful aid in delimiting the boundaries between the soil types. The major confounding factor in analysing the relations between tone and both texture and drainage is the natural colour of the soil. Dark soil tones are characteristic of black, red and dark brown soils, even though these soils may have good drainage properties. Equally, lighter coloured soils occupying poorly-drained areas can be marked by light tones; an excellent example of this is afforded by alkali soils of arid regions which are commonly light grey in colour. It will be apparent from the study of tone, that this property of aerial photographs provides suggestive, rather than conclusive evidence of soil properties. A knowledge of the essential features of soils in the study area, whether by field inspection or by consulting soil memoirs, is a necessary step in order to use the aerial photo to full advantage in soil studies.

Vink (1968) makes a useful distinction between the four basic components of aerial photo interpretation. These are:

1. *Recognition and interpretation* This involves obtaining as complete a picture of ground conditions as is possible from the aerial photograph. Acuteness of observation is the primary pre-requisite.

2. *Analysis* This involves studying the interrelationships of features identified on the photograph. This should be carried out in a systematic manner and will involve the use of overlays.

3. *Deduction* This deals with a combination of careful observation of the aerial photograph and knowledge from other sources in order to obtain information that cannot be directly observed in the photo image. This is where local knowledge may be necessary, as false deductions can lead to gross inaccuracy.

4. *Classification* In this, soil features are grouped into classes, and the final result is a preliminary hypothesis relating to any aspect of the genesis or utilisation of soils.

Whilst the experienced photo-interpreter will combine two or more of these aspects in his work, the beginner is strongly recommended to follow this systematic sequence, making full notes under each heading in turn. In this way, the commonest fault of aerial photo analysis - the examination of the print in a rather aimless, unstructured way - will be avoided.

4.3 PHYSIOGRAPHIC ANALYSIS

The examination of stereo-pairs of aerial photographs permits the description, interpretation and analysis of geomorphological features at the earth's surface. This aspect of studying aerial photographs is most commonly referred to as *physiographic analysis*, and involves the recognition and delineation of landforms. Whilst such a study is central to any work on the geomorphology of a particular area, it is also an important, and perhaps necessary, preliminary in the study of related features such as soils, natural vegetation and land use. In particular, relationships between soils and physiography have been recognised in many parts of the world, and hence play a key role in the study of soil distributions. The use of topographical maps for this purpose is discussed in Section 5.3. The present section is concerned with the value of aerial photographs for studying physiography as a separate component of the physical landscape, although it should be realised that normally it will be a part of a wider soil or land use investigation.

Physiographic analysis can be undertaken at two working levels using two separate scales of aerial photograph. General physiographic or terrain analysis can be used on small-scale photographs to delineate units of the earth's physical surface that have been called either *land facets* or *land units*. Land facets are uniform areas of terrain or, if variable, have definite variations that are predictable wherever that facet is found (Webster, 1962). Webster and Beckett (1970) regard the land facet as a unit from which extrapolations can be made about soil properties from a limited number of field observations. Thus physiographic analysis is important to soil surveyors mapping at reconnaissance scales, on account of the speed, and hence cheapness, at which the ground can be covered.

The examination of micro-scale geomorphological features on large-scale aerial photographs is the second aspect of physiographic study. Although geomorphological features are more usually studied in a composite way by delimiting the entire unit, (for example, an alluvial fan, a solifluction lobe, an alluvial terrace) it is important to realise that these units possess an internal variability that may not be apparent from soil maps. The variability may be of geomorphological or pedological origin. Good examples of the former are given by patterns associated with periglacial patterned ground, alluvial networks in river plains, and estuary creek lineaments in reclaimed coastal marshes. A striking example of the latter is provided by gilgai phenomena in sub-tropical vertisols.

Most areas show such small-scale features, and the main requirement is the procurement of photography at suitable scales. Photographs at a scale of 1:5000 are necessary, and from these it will be possible to recognise any visible relationships between physiography, soils and vegetation. Areal patterns on the micro-scale are, in fact, clearer than is observable by ground investigation.

The production of a map of land facets or land units forms a very instructive exercise, whether as an end in itself, a preliminary to fieldwork or as a preliminary to the study of land use. Photographs with scales of the order of 1:100,000 to 1:20,000 are required, as the use of scales any larger than this will result in broad patterns of physiography being obscured by detail. There is no rigid sequence of operations for producing the map, but the following schedule forms a reliable basis:

1. The area is studied stereoscopically so that the worker may become familiar with the main elements of the relief pattern. Additional aids such as topographical and geological maps should be consulted at this stage.
2. From the preliminary examination, a classification is constructed of the principal land facets found in the area. The terminology used will obviously depend on the nature of the terrain being studied. The actual label given to each land facet is of no great importance, provided that a thorough description is given of each facet. This should include not only details of relief, but also information on surface characteristics, vegetation, land use etc. As the aim of the exercise is to produce a meaningful simplification, most areas will have between four and eight facets. The objectives of the exercise will become lost if the number of facets exceeds this.
3. Having decided on a manageable classification, boundaries can now be drawn on the aerial photograph with a chinagraph pencil, or on a tracing overlay. Benzene will remove all traces of chinagraph without harming the photograph when the exercise is complete.

Once compiled, the map of land facets can provide a basis for a whole series of supplementary studies. If the area can be visited, quantitative information on the classes can be acquired, for example, slope angles, and surface micro-relief. Equally, soils occurring on all or some of the land facets can be studied by sample pits. The relations between geomorphology and soils will thus start to emerge. As a basis for further study, of course, the map will provide a useful tool for correlation exercises with parent materials, drainage and vegetation, and also cultural features such as present land use and historical settlement patterns.

Figure 4.1 and Plate 3 illustrate an example of physiographic analysis on a tract of terrain immediately north of Amman, Jordan.

This distinctly hilly area consists of steep-sided wadis incised into Cretaceous limestone strata. Terra rossa or red mediterranean soils have formed over time, but are now being quickly eroded by hill-wash processes. In many areas the entire soil cover has been removed, leaving bare rock outcrops, both in gently and steeply-sloping sites. Those areas which retain some soil cover are picked out on the photograph by fields supporting dry-farmed cereals. Five land facets have been defined and mapped from the study of the photographs. One example of the importance of this facet map is provided by comparing one facet, e.g. 'moderately sloping soil areas', with the distribution of present arable land. Arable cultivation on this facet is clearly going to lead to severe soil degradation, and such areas would demand a high priority in any scheme of soil conservation.

4.4 CORRELATIONS BETWEEN THE AERIAL PHOTOGRAPH AND THE SOIL MAP

In many countries, the aerial photo is regarded as an essential tool in soil survey and mapping. Most soil maps have, to a greater or lesser extent, been produced with the aid of aerial photos, although the debt which they owe to photo-interpretation varies from area to area. In some cases the analysis of the aerial photograph adds little to large-scale mapping in lowland arable areas (Jarvis, 1962); in other cases the aerial photograph proves extremely important in the final delimitation of soil mapping units (Soil Conservation Service, 1966).

The comparison of a detailed soil map and an aerial photograph is a very useful exercise for: (a) comparing the distribution of soils with the physical and cultural elements of the landscape and (b) assessing the ability of the aerial photograph to distinguish between soil categories on the basis of tonal and textural variations in image.

The first aim is clearly exploiting the property that aerial photographs have of depicting all surface features, whereas all maps are selective in the details that they show. The second aim is more general in scope, in that it will help the student to become familiar with the strengths and weaknesses of the aerial photo as a tool. He is essentially comparing the aerial photo with a map which has been produced by aerial and ground study. Furthermore, by having the mapping units presented to him, his analytical and deductive powers are being channelled to some extent, a feature that many students meeting aerial photographs for the first time find very helpful.

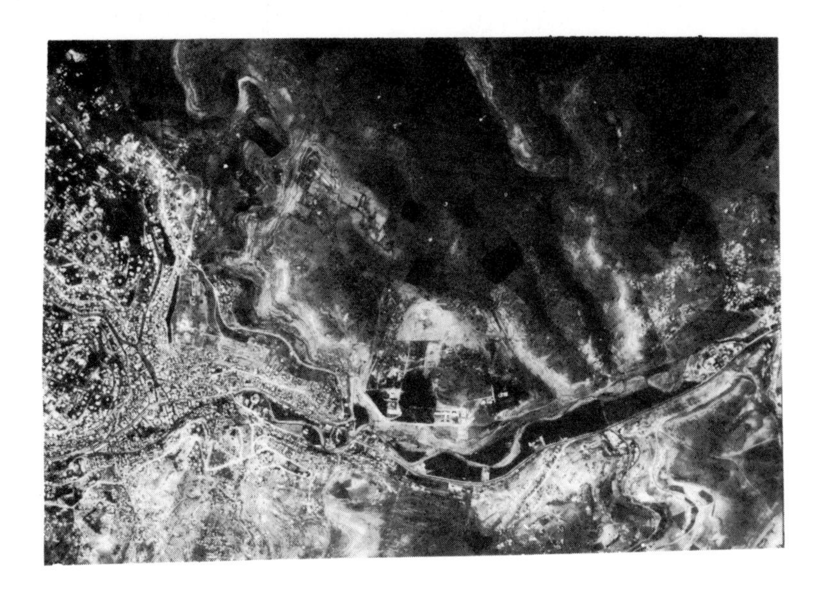

PLATE 3 *The landscape near Amman, Jordan*

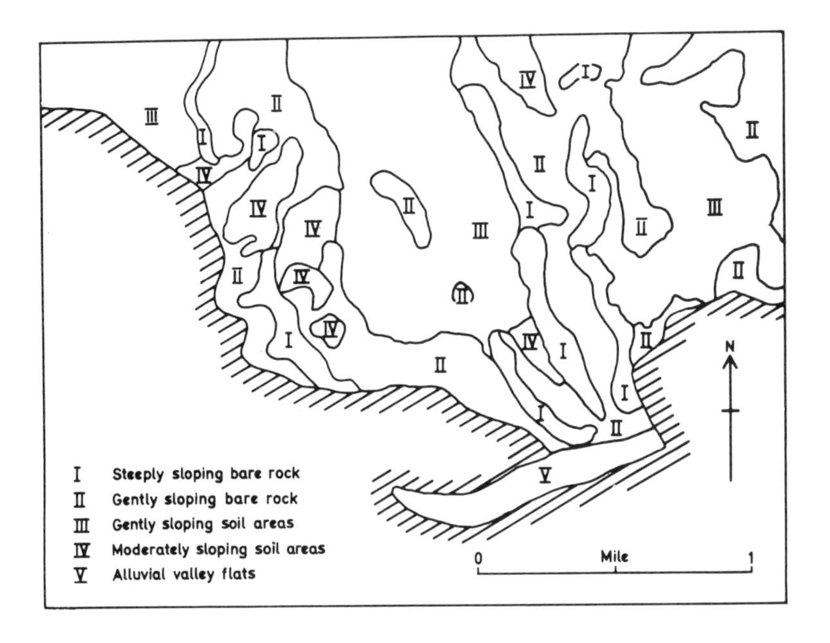

FIGURE 4.1 *Land facets near Amman, Jordan*

Plate 4 shows the result of an exercise to compare the aerial photograph with a soil map. The area depicted lies just to the east of Armthorpe, Yorkshire, England, an arable farming area with flat relief. In this example, soil boundaries have been added to the photograph in ink; normally, however, it is advisable to use chinagraph pencil or an overlay, unless the finished product is to be used for display purposes (see also Table 4.2).

It is important to note that the mapping unit employed is the

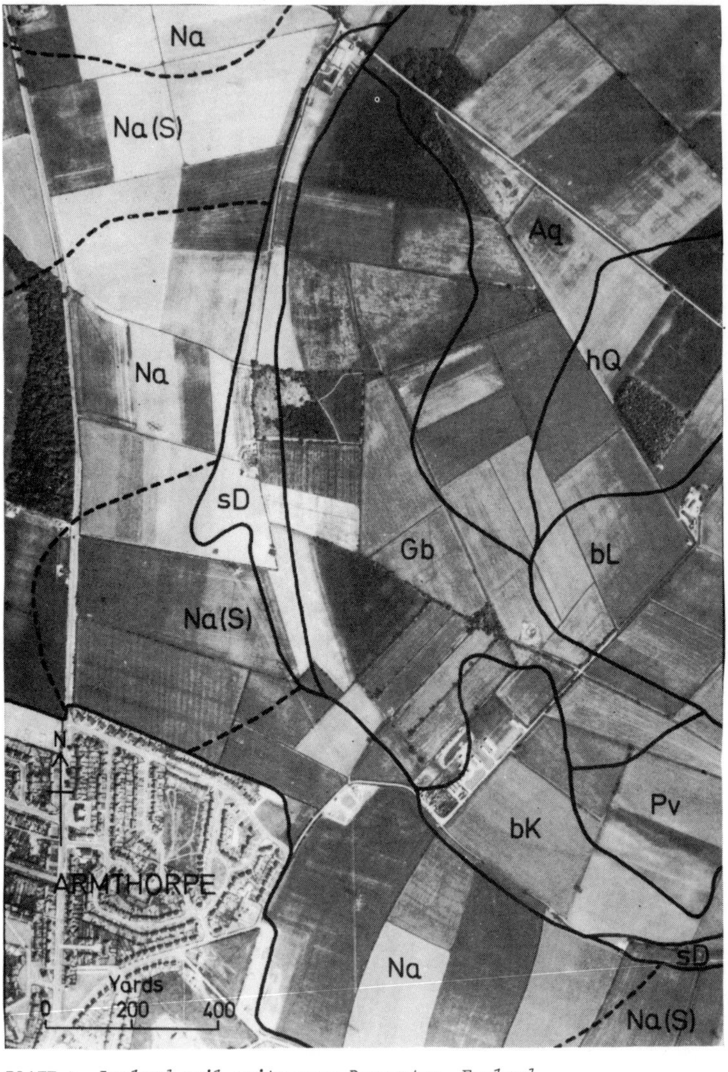

PLATE 4 *Lowland soil units near Doncaster, England*

soil series and, in one case, phases of that series. As an exercise in the cartographic laboratory, the use of the more general soil associations has little to recommend it. The analysis of tone and texture is in general searching enough, without introducing variability within the soil mapping unit.

The requirements for the exercise are a good quality photograph, preferably at a scale of 1:25,000 or larger, a soil map showing soil series and field boundaries, and overlay materials. The aerial photograph need not necessarily be at the same scale as the soil map, for, providing field boundaries are depicted on the map, soil boundaries can be transferred to the aerial photo by interpolation.

SOIL GROUP	LITHOLOGY	SOIL SERIES	SYMBOL
Acid brown earth	Sandy drift (glacifluvial, deltaic)	Newport	Na
		stony phase	Na(S)
Non-calcareous ground-water gley	Sandy drift (glacifluvial)	Stockbridge	sD
	Loamy drift (glacifluvial)	Blackwood	bK
Non-calcareous undifferentiated gley	Clay drift with thin sands (estuarine)	Biggin	bL
	Lacustrine clayey drift with sandy surface	Portington	Pv
Gley Podzol	Sandy drift (glacifluvial)	Gilberdyke	Gb
Peaty gley	Silty drift (estuarine)	Highwater	hQ
Organic	Reedswamp peat (fen-carr)	Altcar	Aq

TABLE 4.2 *Key to accompany Plate 4*

A suitable schedule for the exercise is as follows:

1. Single photographs are studied initially to get an overall view of the area, and then studied stereoscopically for a detailed appreciation of physical and cultural features.
2. The relevant section of the soil map is studied, together with the map legend. No merging or grouping of the soil series

should be attempted; soil complexes, if any, should not be allocated to other series units.

3. Boundaries of the soil units are now transferred to the overlay. Even if the aerial photo and the map are nominally at the same scale, a tracing from the soil map should never be used, due to scale distortions on the photograph. Of all the physical and cultural features that can aid in the process of transference and interpolation, perhaps field boundaries are the most useful. This is one of the main reasons for restricting the exercise to larger scale maps.

4. It is now possible to analyse the relationships between the soil series and their physical context and between the soil series and land use. This is perhaps most easily achieved by making notes in a systematic way, following a scheme such as is shown in Table 4.3.

	Relief	Drainage	Natural vegetation	Land use	Other comments	Photo tone, texture, and pattern
Series A						
Series B						
Series C						

TABLE 4.3 *Check list of soil series characteristics*

Items commented on can be varied to suit the characteristics of the area under study. It is important, though, that a column is always provided for photographic tone and texture. In this way the interpreter is not only interpreting features visible on the photograph, but also continually reviewing the nature of the evidence. This is a necessary discipline, given the need to use the aerial photograph in a cautionary way. Plate 4 provides a good example of this. The geomorphology of this lowland area is quite complex, involving sedimentation in a lagoonal area in the Pleistocene, with a complicated sequence of infilling by glacial, fluvioglacial, lacustrine and estuarine processes. Some aspects of photographic evidence are especially clear. The distinction in *tone* between Portington (Pv) and Blackwood (bK) in the south-east part of the photograph is primarily due to the well-drained status of the latter, giving lighter tones. The distribution of the Stockbridge (sD) is clearly reflected in photographic *pattern*. Similarly, variability within series is shown by variations in photographic *texture*. It is important to realise, however, that photo evidence is never completely universal. Differences between series appear clearly in some fields, but not at all in others. Many things other than soil type affect tone, and lowland arable cropping areas, such as the one shown, present the most difficult problems to the photo interpreter.

4.5 INTERPRETING SOIL CATEGORIES

In non-agricultural areas, physiographic analysis will provide a basic map of land facets. It is often possible, however, to take the study of soils one stage further by directing attention to the landform-vegetation-soil interrelationships of the area. In this way a *soil landscape* map may be constructed which will provide more information on the ecology of the area than is given by geology, physiography or vegetation maps alone. As such, the exercise provides a preliminary soil map at a category higher than soil series level, and is thus a good exercise to attempt before field investigations are carried out, wherever the latter are possible.

The exercise is suitable for any natural or semi-natural area where intensive agricultural management has not obliterated vegetation patterns. Thus areas beyond the limit of agricultural improvement provide good opportunities, for example, where this is due to aridity or to altitude. Areas of high forest are not suitable, due to the obliteration of ground features, although these areas in themselves might provide good exercises on vegetation mapping.

Figure 4.2 and Plate 5 illustrate the results of an exercise of this type carried out on Great Dun Fell, Cumbria, England. The area is part of a National Nature Reserve, and the only agricultural use is for extensive sheep grazing. The amplitude of relief shown in the photograph approaches 2000 feet, and marks a scarp edge to the west through Carboniferous limestones, sandstones and shales. Vegetation is generally of moorland type, with extensive areas of blanket peat. Background physical information is provided by Johnson and Dunham (1963).

Figure 4.2 and Tables 4.4, 4.5 and 4.6 illustrate the

PLATE 5 *Upland landscape near Cross Fell, England*

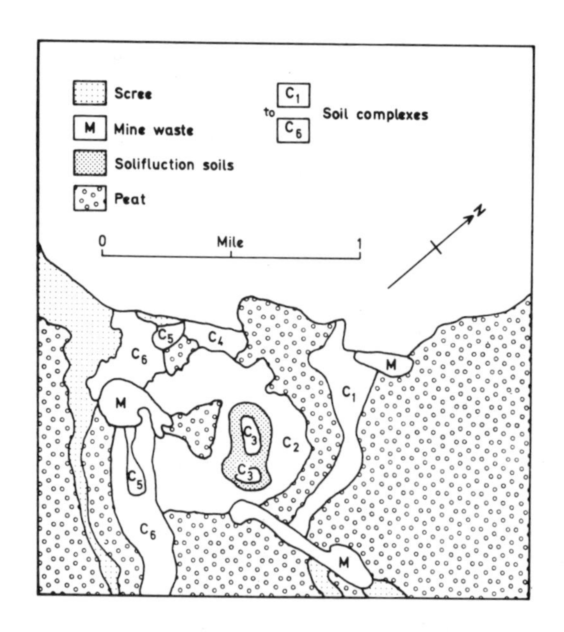

FIGURE 4.2 *Upland soil categories, Cumbria, England*

Slopes	Tone and pattern	Micro-features	Diagnostic properties	Interpretation
Steep	Uniformly light grey	Boulder trains aligned downslope	Surface stoniness and steep slopes	Scree
Flats and slopes	Variegated	Ravines and mounds	Linear, aligned as if along mineral veins	Mine waste
Steep	Uniformly dark grey	None visible	Continuous vegetation cover on steep slopes	Colluvial soils
Flat to moderate	Dominantly very dark	Deeply gullied by linear channels	Tone and erosion evidence	Blanket peat

TABLE 4.4 *Interpretation of soil categories*

sequence of photo interpretation in this area. Table 4.4 summarises the evidence for the four soil categories that can be positively identified. All are identifiable from the photograph, except perhaps for the 'mine waste' category which depends upon a knowledge of the existence of mineral veins in the Carboniferous succession. Table 4.5 presents detail on those categories which can be delimited but not identified. Some further deductions are possible here. C3 is light in tone, indicating good

Slope	Tone and pattern	Micro-features	Diagnostic properties	Designation
Moderate	Complex of light and dark grey	Gullies small and frequent	Variable drainage from site to site	C1
Moderate	Dominantly dark grey	Large gullies occasional	Variable drainage from site to site	C2
Flat	Uniformly light grey	None	Well drained on sandstone	C3
Moderate to steep	Dark grey	Small terracettes	Well drained slopes	C4
Flat to Moderate	Dark and light grey	Sink-holes	Well drained on limestone	C5
Moderate	Variable light and dark grey	Gullies small and frequent	Similar to C1	C6

TABLE 4.5 *Interpretation of soil complexes*

Photo analysis designation	Field checking
C1	Peaty gleys
C2	Peaty gleys
C3	Podzols
C4	Brown earths
C5	Limestone soils
C6	Peaty gley-peaty podzol and peaty podzol-brown earth

TABLE 4.6 *Identification of soil complexes*

soil drainage, and the geological map indicates that it occurs on sandstones; deduction would clearly point to a podzolic type of soil. C5 is marked by sink-holes and, on the geological map, by a narrow limestone outcrop; evidence points to a limestone soil complex. Further deductions from the other soil complexes would be little more than guesses, and the study of profiles in the field or information from existing maps is necessary. The results of this are shown in Table 4.6. As with all photo-interpretation work, it is important to recognise which of the *mappable* categories are also clearly *identifiable*, and the method of presenting information on these should be in an orderly and systematic way.

In working on natural or semi-natural areas such as this, where vegetation plays an important role in the analysis, it is important to note in which *season* of the year the photographs are taken. Most photography is done during the summer months when flying conditions are more likely to be ideal. Careful note must be made of the season of photography and allowances made in constructing the photo key. For example, Plate 5 taken in early summer shows the deciduous purple moor grass, *Molinia*, by dark tones; in late autumn and winter the tone of this vegetation type would be white. Each feature of the landscape undoubtedly has an optimum season for photography and interpretation; however, photographic coverage usually has to serve several purposes and one must thus pay special attention to season in compiling the key.

4.6 STUDYING SOIL EROSION FROM AERIAL PHOTOGRAPHS

Maps showing the extent of soil erosion in a particular area provide useful additions to those of topography, land use and soils. Field investigations and checking provide vital information on rates of erosion, but soil erosion can often be identified directly from the photograph. This is an important interpretive advantage of using aerial photographs in erosion surveys, as in many instances the evidence is clearly visible, e.g. deep gullies in arable fields. In other cases processes of erosion have to be deduced from visible evidence, e.g. sheet erosion as evidenced by stoniness in residual soils. Thus, it is often possible to make a preliminary assessment of soil erosion, at a general level, of areas not immediately familiar to the interpreter. Therefore exercises on soil erosion assessment can be based on classic areas, e.g. Mediterranean lands, semi-arid United States or inter-tropical Africa where soil erosion processes and problems are quite well documented. If photography is available for a reasonable time interval, e.g. ten years or more, it is possible to carry out more refined work on rates of erosion.

The following kinds of erosion can normally be distinguished, given good quality photo cover at scales larger than 1:25,000:

1. Stream channel erosion, i.e. valley form
2. Stream bank erosion, i.e. slumping of banks
3. Gully erosion, i.e. valley trenching by concentrated water flow
4. Rill erosion, i.e. incipient and small gullies
5. Sheet erosion, i.e. topsoil removal by continuous sheet floods
6. Severe wind erosion, i.e. wind-scoured and deflated surfaces
7. Spectacular and catastrophic events, i.e. slides, slumps, mud flows, flood scours

The absence of signs of the above processes does not mean that no erosion is taking place, as some types of sheet erosion are not detectable from the photograph and soil creep too is difficult to recognise.

In addition to processes of erosion, the aerial photograph will clearly show what measures of soil conservation, if any, are being undertaken at the present time. Such measures will include engineering control (e.g. dams and terraces), field cultivations (e.g. contour ploughing and lister ploughing) and cropping patterns (e.g. strip cropping). Where there appears to be no evidence of conservation measures, it becomes an instructive exercise to make a preliminary assessment of what measures seem to be required both at regional (i.e. catchment area) and local (i.e. farm) level. In practice, soil conservation measures are extremely complicated to design and lay-out, but at a general and reconnaissance level it is possible from the aerial photograph to discuss what types of control are indicated in different erosion areas. This will involve a discussion of types and causes of erosion, and will provide a good test of the student's background knowledge.

Figure 4.3 and Plate 6 show the results of a soil erosion exercise applied to part of Adams County, Nebraska, U.S.A. The landscape shown is composed of the loessial silts so characteristic of the Great Plains. Three major landform elements are clearly visible from the aerial photo, namely, general undulating swells in relief, drainage patterns, and erosional features. In this area, ground swells are relatively subdued and are about half a mile in breadth. The drainage pattern is strikingly dendritic, with a tendency for side channels to be straight and parallel. Erosional features are most clearly seen along the drainage lines and take the form of steep, almost vertical, stream banks. Structurally, loess tends to cleave vertically and often exhibits near-vertical slopes. Thus gully channels formed in loess have low gradients and steep sides, often highlighted by slips and terracettes.

The regional soil profile is the mollisol (brown chernozem to dark-brown chestnut) which shows generally greyish tones, although these are considerably influenced by the cropping pattern in the area. In contrast to the darkish-coloured mollisol surface, loess is naturally grey to yellow in colour. Thus its photographic tones are characteristically light where there is no continuous vegetation cover. In addition, as loess is well drained, the light tones are rarely darkened by moisture.

Tonal contrasts between light grey and grey are important in the example being studied. Light areas particularly along gully sides and forming mottled patches in some fields, delimit areas where erosion of organic surface horizons is taking place; thus one can infer sheet erosion or wind erosion in areas where no gullies are visible. An additional hypothesis can be suggested, namely, that the lightness of tone is proportional to the severity of removal by sheet-erosion. This would obviously require field testing, but it does provide a basis for classifying the area into erosion categories. Whilst areas of extreme and slight erosion can be demarcated reasonably confidently, the intermediate intensities are less easy to delimit. The working hypothesis suggested above is of great help in these middle categories.

A suitable schedule for studying soil erosion from aerial photographs is as follows:

1. The area is studied stereoscopically and notes made on visible physical and cultural characteristics. Background documentary material can be consulted at this stage.

2. Soil erosion and soil-surface conditions are now examined as closely as scale permits. Observations and deductions

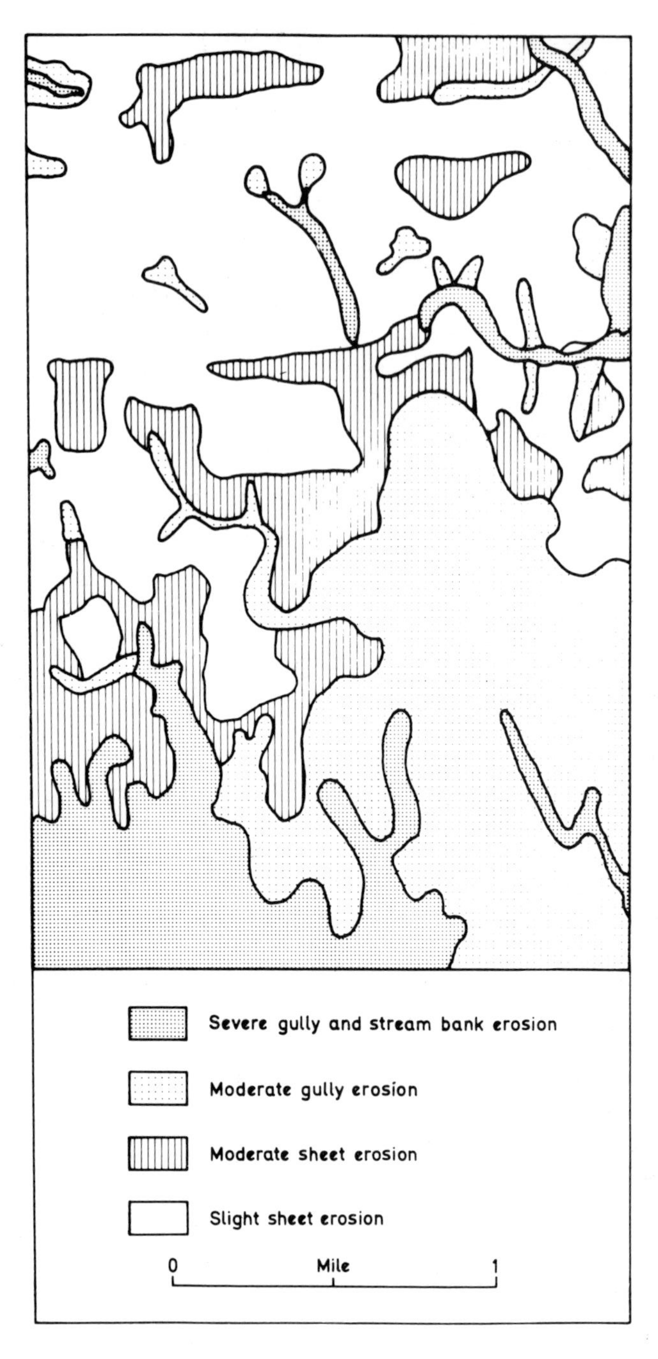

FIGURE 4.3 *Categories of soil erosion, Nebraska, U.S.A.*

PLATE 6 *Soil erosion patterns in Nebraska, U.S.A.*

are recorded.

3. A classification of erosion types is now constructed, and an overlay made from the aerial photograph.
4. Any soil conservation measures being employed in the area are marked on the photo or on a separate overlay.
5. Recommendations are listed with regard to the extension of present conservation measures or the introduction of new ones.

The classification of the study area into sub-areas on the basis of erosion type and intensity will largely depend on the nature of the region and the scale of mapping. Various combinations can be used according to the needs of the particular case. A detailed analysis (Vink, 1968) might include the following categories:

GG	Severe gully erosion
GGP1	Potential gully erosion, with severe sheet erosion
GGP2	Potential gully erosion, with moderate sheet erosion
GGP3	Potential gully erosion, with slight sheet erosion
SG1	Sheet and gully erosion severe
SG2	Sheet and gully erosion moderate
SG3	Sheet and gully erosion slight
SS	Severe sheet erosion
S	Moderate sheet erosion
(S)	Slight sheet erosion

Normally, however, a scale as refined as this will only result from a complete soil erosion survey involving detailed ground control. From three to six categories will normally be adequate from the study of an aerial photograph.

REFERENCES

American Society of Photogrammetry. *Manual of photographic interpretation*, Washington, D.C., 1960.

Ball, D.F., Hornung, M. and Mew, G. 'The use of aerial photography in the study of geomorphology and soils of upland areas', in *The application of aerial photography to the work of the Nature Conservancy*, Edinburgh, 1971, pp. 66-77.

Clarke, G.R. *The study of soil in the field*, (5th Edn.) Clarendon Press, Oxford, 1971.

Dickinson, G.C. *Maps and air photographs*, Arnold, London, 1969.

Evans, R. 'Air photographs for soil survey in lowland England: soil patterns', *Photogramm. Record*, 7 (1972), 302-22.

Hunter, G.T. and Bird, S.J.G. 'Critical terrain analysis', *Photogramm. Engng.*, 36 (1970), 939-55.

Jarvis, R.A. 'The use of photo-interpretation for detailed soil mapping', *Arch. Int. de Photogramm.*, 14 (1962), 177-82.

Jenny, H. *Factors of soil formation*, McGraw Hill, New York, 1941.

Johnson, G.A.L. and Dunham, K.C. *The geology of Moor House*, H.M.S.O., London, 1963.

Kodak Ltd. *Kodak data for aerial photography,* Rochester, U.S.A., 1969.

Mott, P.G. 'Some aspects of colour and aerial photography in practice and its applications', *Photogramm. Record*, 5 (1966), 221-37.

Soil Conservation Service. *Aerial-photo interpretation in classifying and mapping soils*, Agriculture Handbook 294, Washington, D.C., 1966.

Tarkington, R.F. and Sorem, A.L. 'Colour and false colour films for aerial photography', *Photogramm. Engng.*, 29 (1963), 85-95.

Vink, A.P.A. 'Aerial photographs and the soil sciences', in *Aerial surveys and integrated studies*, UNESCO, Paris, 1968, 81-141.

Webster, R. 'The use of basic physiographic units in air photo interpretation', *Arch. Int. de Photogramm.*, 14 (1962), 143-48.

Webster, R. and Beckett, P.H.T. 'Terrain classification and evaluation using air photography: a review of recent work at Oxford', *Photogrammetria*, 26 (1970), 51-75.

5 SOIL MAP INTERPRETATION

5.1 VARIETY AND PROPERTIES OF SOIL MAPS

The soil map is a generalisation, in cartographic form, of the areal distribution of soil types over any defined part of the earth's surface. Although many areas of the world are now covered by soil maps, there is no universally agreed basis for the definition of mapping units since local differences of soil and terrain often demand independent approaches to classification and mapping. National and international soil survey organisations have different aims and different priorities and it is almost inevitable that different groupings and nomenclature will be found from one organisation to another. Some organisations which supply soil maps are listed in Appendix A. It will often be wise to consult the primary publication, whether it be a soil report, special publication or even a published article, which explains the rationale of the system. A selection of such guides is included in the bibliography and the items marked by an asterisk.

The most important aspect of any soil map is the initial decision on the mapping unit to be employed. The scale at which a map is to be published determines both the scale at which the soil surveyor maps his soils and the criteria used to differentiate the units. For this reason, maps produced on national scales (e.g. 1:1 million to 1:10 million) can generally do no more than show the distribution of major soil groups or complexes of major soil groups, whereas maps at large scale (e.g. 1:10,000 and less) are able to portray individual soil properties. Thus the information content of large-scale maps allows a detailed resolution of the ground pattern of soils, although it is appreciated that soil maps on different scales are generally prepared to meet different requirements. The map legend reflects these differences and will determine the usefulness of the map.

Duchaufour (1970) has divided soil maps into three convenient classes, as follows:

a) Small scale 1:1,000,000 to 1:250,000
b) Medium scale 1: 100,000 to 1: 50,000
c) Large scale 1: 20,000 to 1: 5,000

Small-scale soil maps are, in practice, less maps of soil than of major soil-forming environments. For example, the influence of climatic regions is illustrated by the continental soil maps of North America and Africa which show major soil groups. Similarly, in interior Australia where rainfall is uniformly low over large areas, soils have been grouped according to their association with major landform systems. Medium-scale soil maps, on which individual farms, roads and urban areas can be located, depict local variations in soil profile characteristics. These variations can often be identified within a single farm and be used in rural land use planning. Soil associations are usually shown on maps in the scale range 1:100,000 to 1:250,000, and soil series can be shown on maps from 1:100,000 to 1:50,000. Large-scale maps are a practicable means for resolving detailed questions on the use of soils for specific purposes, for example the design, layout and management of projects concerned with agricultural experimentation, drainage construction, irrigation provision and urban uses. For this reason, a crucial scale threshold for the utility of any soil map, is whether it portrays individual fields, the basic management unit for many land use purposes. Not only can more detail be shown, but also the scale allows the distribution of individual soil properties to be given (e.g. texture, permeability, soluble salt content), as distinct from genetic mapping units which aggregate the properties of whole profiles.

5.2 RELATIONSHIPS OF SOIL TO SOLID AND SUPERFICIAL PARENT MATERIALS

Although even the pattern of world zonal soils is not without the influence of distinctive parent materials, investigations of the kind described here fall within the province of the medium and large-scale soil maps which show genetic soil types usually subdivided into series or associations. In many instances the factors limiting this investigation in a given locality will be the availability of solid or drift geological maps together with a soil map preferably on the same scale. A further constraint affects the manner of presentation. If it is desired to annotate a geological cross-section with soil details it must be appreciated that any vertical exaggeration to emphasise variations in relief (see Section 5.3) will destroy the true thickness and dips of strata. Many geological survey maps are nevertheless accompanied by a sectional representation often with a vertical scale about three times that of the horizontal. The examples given below are of geological and soil transects which normally utilise tracing overlays but are presented here in Figures 5.1 and 5.2 as separate pairs of maps.

The chosen transect need be no larger than about 2 × 4 inches but should possess internal variety of soil parent materials. For demonstration purposes any maps will normally have certain areas over which correlations seem clearer, but experience shows that in this, as in the exercises which follow, it is often best for individuals to choose their own patch of territory. As a result, individual satisfaction will vary and yet group work and ensuing discussion provides a good test of the workability of the method. Essentially, tracings of parent materials and soils are to be compared and the success of this venture depends upon how imaginatively the respective mapping legends are used. For example, precise lithologies rather than age-names are important on the geological map since texture, permeability, base status and other attributes are all primarily consequences of lithology and may in themselves provide a more precise comparison than coarser categories such as major soil groups. When pedogenesis is initiated, the texture and mineralogy of the parent material plays a dominant role in the evolution of the profile. The mineralogy and basicity will largely determine the weathering reactions that take place within the soil body and the texture in turn will influence moisture and air movement. The significance of the soil series in this context is that each series is related to a single parent material and a particular gradient and position in the landscape, while the association is a grouping of all the series which occur on a given parent material.

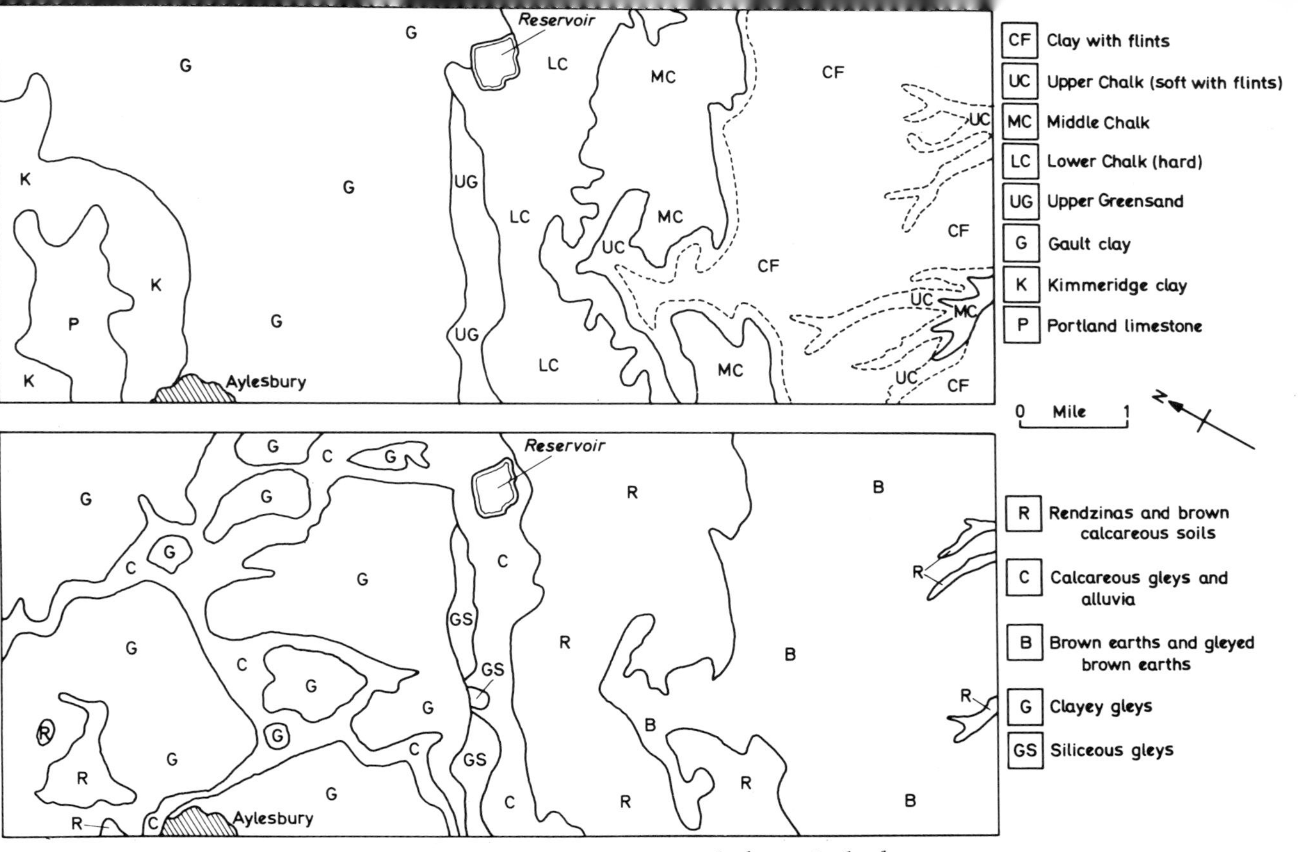

FIGURE 5.1 *Parent material and soil categories, Aylesbury, England*

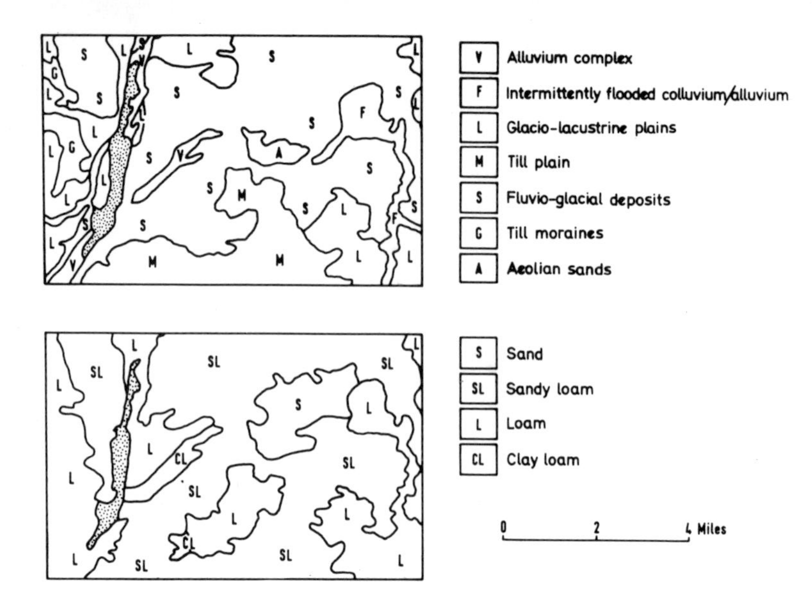

FIGURE 5.2 *Superficial deposits and soil textures, Saskatchewan, Canada*

The method of comparing two tracings is perhaps so simple that it merits little comment and yet this exercise can be fraught with either erroneous or shallow comparisons. Apart from the care needed to obtain two perfectly overlapping transects and the patience required to achieve a faithful portrayal of lines, the different source maps may have slightly different basic grids and may, if produced many years apart, have differing interpretations of the parent lithologies. Again, treatments which aim to compare solid geology with soils, without consideration of the role of superficial deposits (if present), will be as suspect as those which deal purely with the superficial spread of soil parent materials without a thought for the nature of the substratum and its possible influence on drainage. Finally, although textures can always be linked to geological deposits, soil drainage cannot, and must always be considered within a wider topographic context.

The first example is taken from the area astride the escarpment of the Cretaceous Chalk near Aylesbury, England (Sheet 238). The geological extract shows the eight main lithological units with a north-west facing escarpment formed in the Middle Chalk.

A clay vale lies to the north, while dry valleys occur on the dip slope to the south-east.

The soils of the limestone slopes are of rendzina or brown calcareous affinity while those of the chalk dip-slope are predominantly brown earths and their gleyed variants, having an acidic reaction owing to development on the overlying 'Clay with Flints' formation. Part of the Lower Chalk outcrop has been covered by 'head' deposits of Pleistocene age derived from the Chalk upland and these are predominantly of impeded drainage. The clay vale soils are gleyed without exception, and sinuous belts of riverine materials stand out clearly. The Greensand outcrop fronting the escarpment is situated at a natural spring line and its soils are again gleyed but have the distinctive texture imparted by the sand.

The second example shows how superficial parent materials can be related to a specific soil property. Superficial deposits form the parent materials of such a high proportion of the earth's surface that the study of soil characteristics in relation to them is an essential exercise. Figure 5.2 shows a twenty-four square mile extract of superficial deposits and soil textures from the 1:125,000 soil map of the Rosetown Area, Saskatchewan, Canada (Sheet 72-O). This tract of prairie was covered by the Wisconsin glaciation, which has left an extremely complex suite of deposits. In addition to till moraines and till plains, lacustrine deposits and fluvio-glacial materials cover large areas and have been the source of late-glacial aeolian sands. Alluvial complexes form the main post-glacial soil parent materials. In this area, as in many others, the only way to rationalise such a complex mosaic of superficial materials is by understanding the sequence of Pleistocene events.

The soil pattern that has evolved, closely reflects the contrasting parent materials. The dominant regional soil, the chernozem, is found on medium textured deposits, the loam moraines, loamy lacustrine plains and loamy fluvio-glacial deposits. The brown chernozem with lower organic matter content and lower moisture capacity is found on sandy loams. Regosols are formed on the coarse aeolian sands, while areas of clay loam carry brown solonetz soils. The relations between texture and major soil group are very striking. However, the boundaries of superficial deposits and major soil groups are not coincident, a situation which is due to the fact that interpretations of landforms are essentially genetic, with origin of major interest rather than any specific property such as texture. Thus, in the glaciated area, the primary stage is to abstract information on soil textures and establish relations between textures and deposits. A second stage is to compare textural categories with the distribution of major soil groups.

5.3 RELATIONSHIPS OF SOIL TO RELIEF AND DRAINAGE

These correlations again lend themselves mainly to treatment at medium and large scales. There are two main approaches. The first gives an areal representation of soil and slope units while the second gives a sectional representation of soils and slopes along one or more traverses.

For the first method a small transect is selected which embodies appreciable changes of slope. A topographic map will usually be needed for this exercise. A piece of tracing paper is placed over the selected area and a grid of closely-spaced squares is ruled across it e.g. $\frac{1}{4}$ inch when the map scale is 1 inch to 1 mile. The mean slope within each square is then calculated with reference to the following formula, in which θ is the average gradient within each grid square.

$$\theta = \tan \frac{VI}{HE} \frac{(\text{no. of contours crossing square} \times \text{contour interval})}{(\text{distance represented by one side of grid square})}$$

The vertical interval (VI) and horizontal equivalent (HE) must be expressed in the same units, θ being found from mathematical tables. Squares of mean slope are grouped into about six categories and an isopleth map is produced on which shading is intensified with steepness of slope. A quicker alternative method to obtaining mean slope values is to make an isopleth map of the relative relief (relief amplitude) within each square, in which case the isoplethed values represent the difference between the lowest and highest contours. Neither of these methods is very precise but the error involved is standard from square to square and, as it is based on contour intervals, the calculations are greatly facilitated. The following example illustrates the mean slope technique in relation to soils in the Mendip district of Somerset, England (Sheet 280). The object of the exercise is to demonstrate the extent to which it is possible to see a direct relationship between average slope angles and soil character. For this reason emphasis should be placed upon soil drainage status in the selection of soil categories. These can often be simplified from the groupings in the map legend.

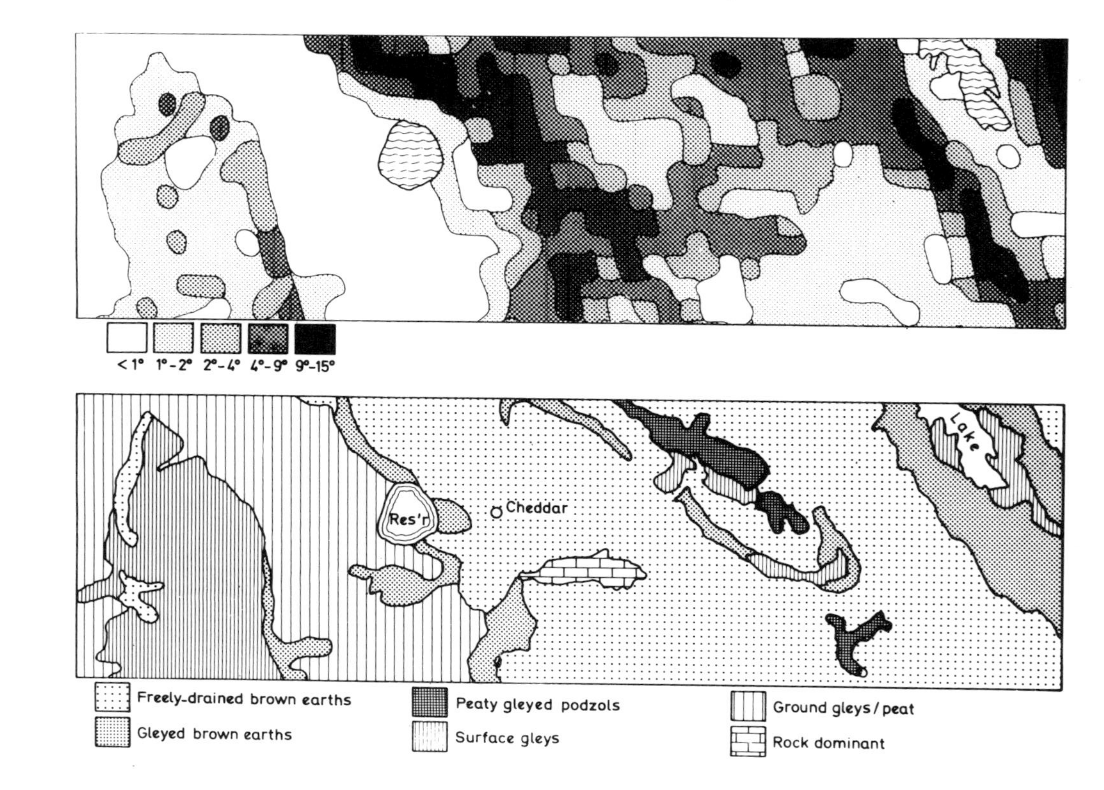

FIGURE 5.3 *Analysis of slopes and soils, Somerset, England*

Selection of suitable areas on the soil map is important but the educational value of this type of exercise, apart from demonstrating what had already been anticipated, lies in attempting to explain residual soil variations which are not directly, nor perhaps to any degree, attributable to topography. As such, it is a tool for isolating the effects of a single soil forming factor.

Figure 5.3 shows that freely-drained brown earths occupy the Mendip slopes and plateau, with gleyed brown earths peripheral to these areas either at the foot of slopes or towards the centre of the Mendip interfluve. Surface water gleys and peaty podzols occur on the interfluve on low-angle slopes, while surface gleys are again seen in the west, though their presence here is attributable to the occurrence of clays and marls. Ground water gleys occupy the extensive alluvial and marshy tracts in the west which were once arms of the sea. Although the Mendip plateau is underlain by hard limestones, it is surmounted in the north by sandstone, while to the south of this there is a variable cover of loess. These two parent materials have facilitated the development of podzolic soils in isolated areas.

The second method of analysing soils and relief involves construction of topographic profiles and their annotation with soil detail. Although some soil maps do show contours they are invariably so overprinted that it is necessary to derive contour detail from topographical maps. For the construction of a slope profile the horizontal scale must be the same as that of the soil map while the vertical scale may often be exaggerated e.g. to five times horizontal scale, in order for the relief variations to be emphasised (Monkhouse and Wilkinson, 1964). For example,a profile with a horizontal scale of 1:25,000 should have a vertical scale of about 1:5000 (1 inch representing 417 feet).

For the single profile or section, a strip of graph paper with axes already drawn, is placed with its upper edge along the line of the traverse. All contour lines which cross the traverse are then marked at their appropriate heights on the section with a fine pencil point. When complete, the surface configuration is drawn in (see Figure 5.4). A variation on this theme is the construction of a multiple section as illustrated in Figure 5.5. Though more laborious to construct, this is suitable for group classwork projects and overcomes to some extent a major weakness of single sections, i.e. that they may not necessarily provide a representative picture of the local soils.

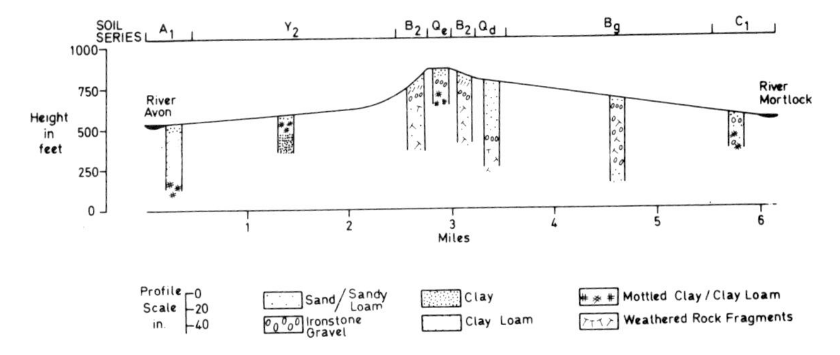

FIGURE 5.4 *A soil and slope analysis, Western Australia*

For the multiple section, a large sheet of graph paper is taken and the proposed traverse lines are marked on it directly from the map. An illuminated tracing table is helpful for this purpose. Axes are again placed at the ends of the traverse lines and must be parallel with the N-S grid lines of the map. This will distort the graph paper squares into rhombuses and result in an impression of perspective, though the result cannot be a true perspective diagram. Judicious orientation of the traverses can, of course, help to achieve a pleasing end product. It is necessary to mark a height scale across each section in order to facilitate correct location of contour points. This is so because one cannot now use the graph paper squares unless a traverse is orientated precisely E-W. Although it is possible to work from one sheet of graph paper it will prove easier to draw each section separately, ensuring that each has the correct amount of angular distortion otherwise they will fail to weld together.

Soil detail is now added to the sections. Several methods exist for doing this, the choice being determined partly by preference and partly by the amount of detail that it is necessary to show. Two methods are illustrated below.

The first example is taken from a 1:63,360 map of the York-Quairading area, Western Australia (C.S.I.R.O. Soils Publ. No. 17) and is illustrated in Figure 5.4. A topographic section is first constructed from relief information given on the soil map and in this example a 6-mile transect has been used. The boundaries of the soil units - in this case soil series - are then added to

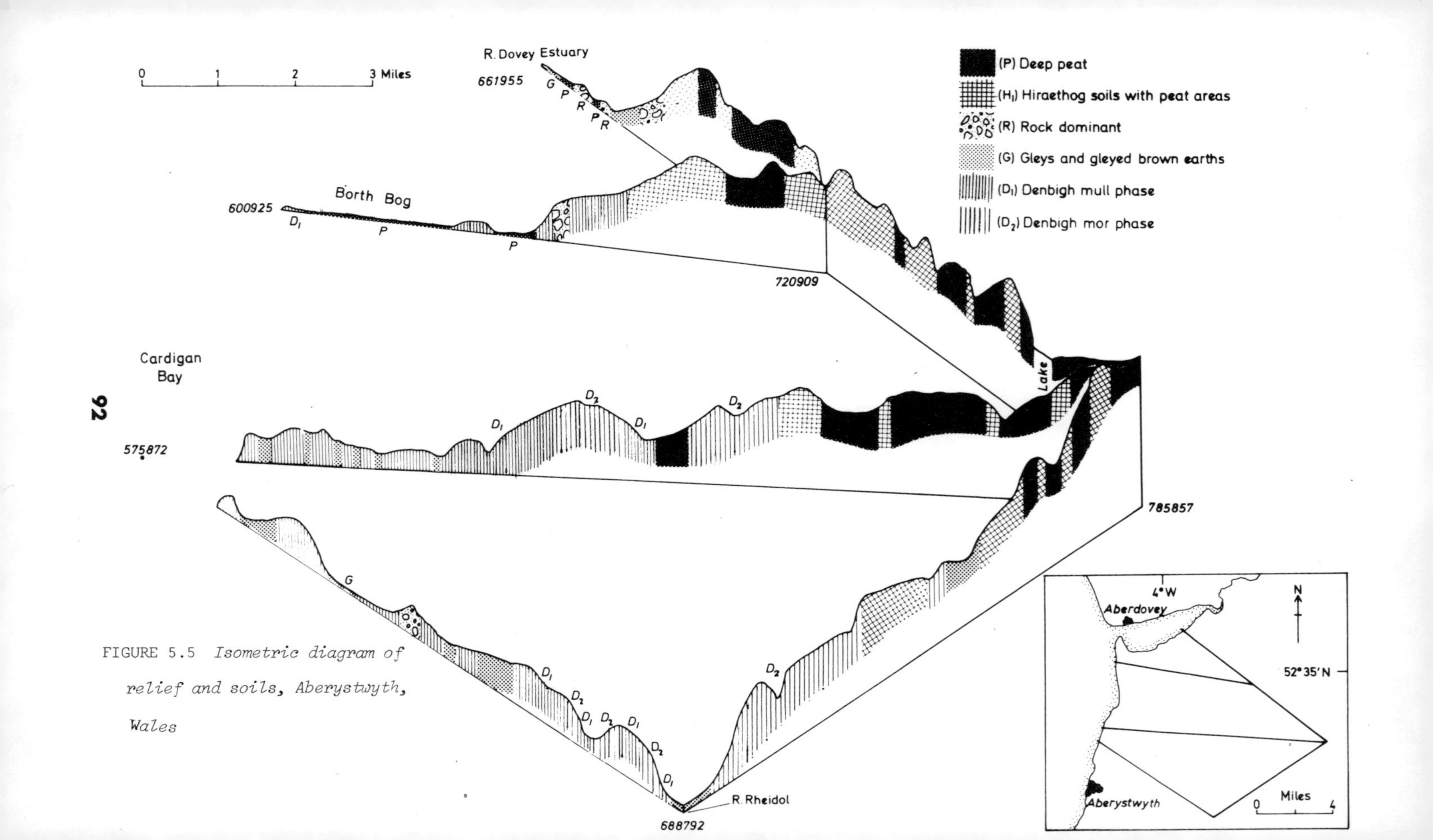

FIGURE 5.5 *Isometric diagram of relief and soils, Aberystwyth, Wales*

the diagram. The final stage of the exercise is to superimpose on the section diagrammatic representations of the profiles of each soil unit. A study of the relevant soil memoir will quickly reveal the key soil horizon characteristics in the area and hence the horizons which need to be shown on the diagrammatic profiles. Sometimes these will be important textural features e.g. sand, clay, clay loam, and at other times they will be pedogenic characteristics, e.g. mottling or ironstone gravel formation. The choice of these key profile features or diagnostic horizons will clearly depend on the kind of area under study. The number should, however, never be larger than about eight, if the diagram is not to become too complex. After choosing the features to be depicted, a symbol is decided upon for each. Duchaufour (1970) gives a list of possible horizon symbols, although it is often quite satisfying to devise one's own. Profiles of uniform width are then drawn on the topographic section in the middle of the particular soil unit and the drawing is made to scale to preserve relative soil depths and horizon thicknesses. After consulting the soil descriptions in the memoir, the details of profile character can now be added to the diagram. Composite symbols are often effective, as in the case of the C_1 series where mottling, sands and ironstone gravel all occur in the subsoil.

Profile details must now be related to topographic situation. In Figure 5.4 it will be noticed that the sequence from Qe to C_1 represents a slope sequence characteristic of arid tropical terrain with the high level residual latosol (Qe) passing into deeper colluvial laterites (B_2, Qd) at the foot of a breakaway. Proceeding from the colluvial zone along the main pediment, residual ironstone soils are found (Bg) which finally give way to alluvial C_1 soils in the Mortlock valley. The position of the soils in this slope sequence can largely be explained by present processes of transport (water, mineral materials, iron, soluble salts) along the slope.

The sequence from Qe to A_1 is less amenable to such an explanation. Whilst contemporary slope processes can explain the main details of distribution, e.g. residual latosol on the shedding site (Qe) and hydromorphic solonetz on the receiving site (A_1), many profile details are related to the fact that the land surfaces - Quailing (Qe), Balkuling (B_2), York (Y_2) and Avon (A_2) - are of successively younger age (see Section 5.4).

The second example (Figure 5.5) is from Wales and comprises

a series of sections from the Central Welsh Uplands down towards Cardigan Bay and the estuary of the river Dovey (Sheet 163). This area has been chosen because of its uniformity of soil parent materials, allowing one to examine local drainage variation as well as the altitudinal impress of climate. Although much reduced from the original 1:63,360 scale Figure 5.5 demonstrates the altitudinal separation of peats and peaty podzolic soils from the brown earth suite below 800-1000 feet. It illustrates some of the characteristic hydrological contrasts between peaty terrain and the better-drained sites on which podzolic soils predominate. It also shows the local distinction which has been made between brown earths of lowland (mull) and upland (mor) affinity, indicating that in these transitional environments biological activity is delicately balanced and may be enhanced or retarded by comparatively small differences in elevation, slope, and certainly by management. The coastal plain possesses a mosaic of drainage conditions, with gleys in depressions which frequently occupy very narrow tracts.

In broad terms the clinosequence from coast to interior is a climatic one, since rainfall increases from 40 inches in the west to around 100 inches in the hills. However, the acidity of the soils and widespread peat accumulation is due also to the non-calcareous nature of the parent materials derived from shales, mudstones and greywackes. This sequence is not to be regarded as a 'regional catena', as soil-slope sequences must be identified separately within a specific altitudinal-climatic zone if one wishes to study the topographic component without climatic overlay.

5.4 RELATIONSHIPS OF SOILS TO VEGETATION AND AGE

There is an almost universal inclination to interpret soil morphology in relation to the environment with which it coexists at the present time. This is understandable enough, yet besides parent material and siting, this environment includes vegetation which is known to have undergone changes, and soils themselves may have been developing for different lengths of time.

Relationships between soil and vegetation can only be studied adequately if there is overlapping map coverage at the same scale. There are many areas where ecological and soil surveys have been carried out, for example, of individual forest areas and other localities of scientific interest such as nature reserves and national parks. Research papers have also illustrated such relationships but usually for small sample areas.

Mapping exercises to investigate vegetation and soil relationships can be approached in two ways. Figure 5.6 shows soil and vegetation maps for a part of Clashindarroch Forest, Aberdeenshire, Scotland. The forest area had not been planted at the time of the survey (Muir and Fraser, 1940) and both maps were produced independently of one another in order to achieve a more objective correlation. The number of vegetation categories has been simplified from fifteen to six and the number of soil categories from sixteen to five. This simplification is necessary if the mind is not to become bewildered by the diversity of categories, and in itself will constitute a beneficial preliminary exercise.

Soil type	Dominant vegetation	Associated vegetation
Peats	*Calluna-Eriophorum* moor	---
Brown earths	*Calluna*	Flush, wet heath
Podzols	Wet heath, *Calluna*, Eroded *Calluna*	---
Peaty podzols Peaty gley podzols	Wet heath	*Calluna*, eroded *Calluna*
Gleys	*Calluna-Eriophorum* moor, wet heath, *Calluna*	Flush

TABLE 5.1 *Systematisation of soil type and vegetation assemblage*

The first approach towards this correlation is to compare tracing overlays based on an area of about 6 square miles. After studying the patterns and arriving at preliminary hypotheses, it is often a good discipline to express this in the form of a table, as in Table 5.1. From the data in Table 5.1 it can be appreciated that whilst peat soils correlate closely with *Calluna-Eriophorum* moorland, all other soil groups are associated with several vegetation types. The low degree of correspondence between soil and vegetation boundaries, in addition to reflecting differences in nomenclature and classification on both sides, points to the importance of non-edaphic factors in controlling vegetation distributions. These factors include local climate, aspect and exposure, dynamism in the plant

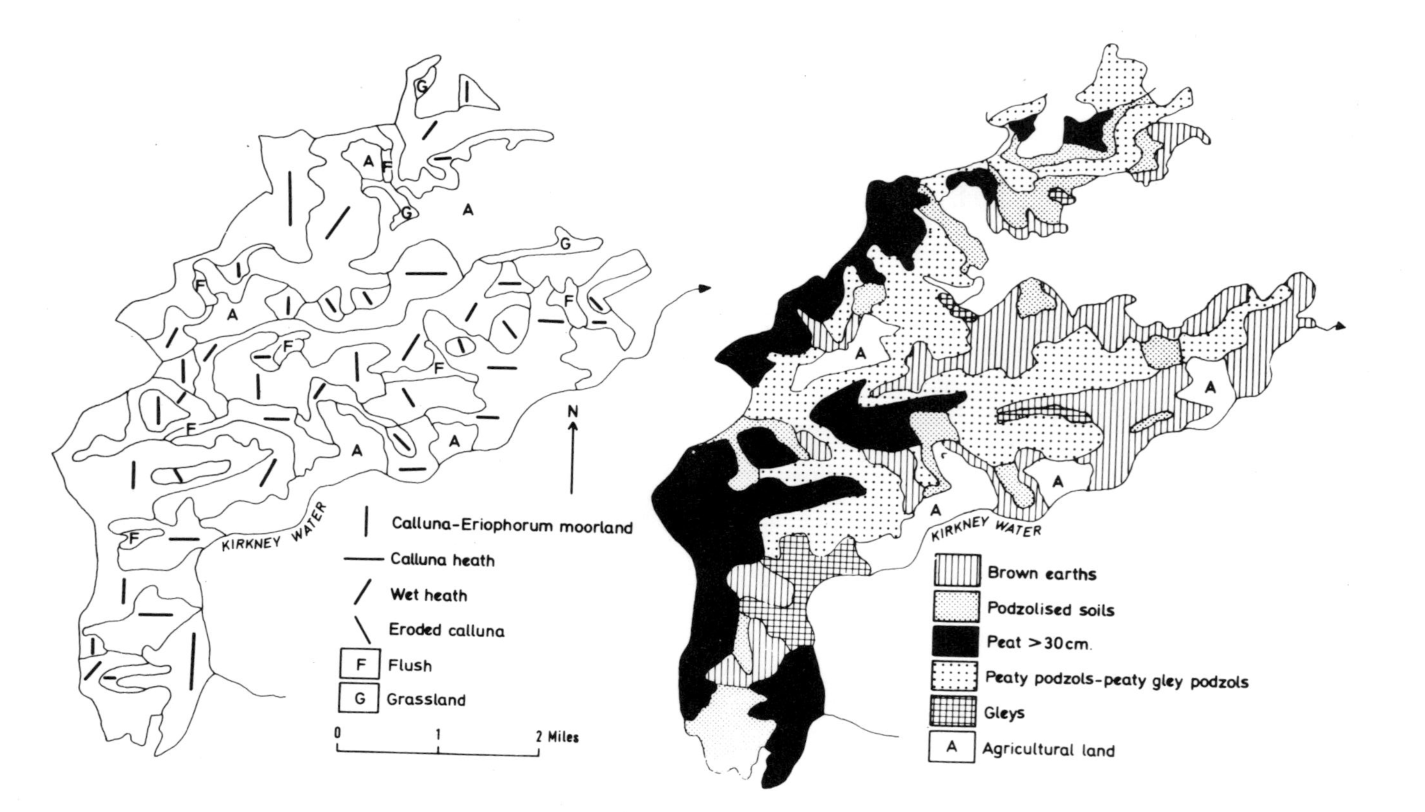

FIGURE 5.6 *Vegetation and soil categories of part of Aberdeenshire, Scotland*

communities and the various facets of human interference. In another locality the relationship between vegetation and lithology might be particularly well marked. Again, the value of this exercise lies as much in the unexplained residuals as in the immediately perceived relationships.

A second and complementary method of studying these relationships is to construct a slope profile, as in Section 5.3, and to superimpose on this the vegetation boundaries above the profile and the soil boundaries on or below it. By this method, detailed categories can more easily be lifted straight from the maps and, of course, topography can feature in the analysis.

Opportunities to study the effect of time on the evolution of soils are restricted because of the many factors which are so continuously changing in space. Hence on a global scale, climatically-directed processes tend to obscure our conception of the widely differing lengths of time that soils have been forming. On a global scale, this point is a particularly important one since the areas of most intense chemical weathering at the present time are also among the world's oldest landscapes, while the slow rates of pedogenesis characteristic of high latitudes (and altitudes) are being conducted on terrain which has only been exposed since the withdrawal of glacial conditions. Even in more restricted localities offering chronosequences of soils, for instance on river or lakeshore terraces, one can rarely be sure that parent materials were initially uniform on all surfaces, and in order to study the time factor, all other variables affecting soil formation should ideally be kept constant (Franzmeier and Whiteside, 1963). There are, for this reason, many more morphological candidates for chronosequence work than *bona fide* examples. Thus local reworking of a particular surface or slight textural variations between adjacent members may account for differences in profile morphology, independently of a distinction on the grounds of age. Nevertheless, given a knowledge of the relative ages of soils, details of their morphologies can often yield some evidence of their evolutionary pathways.

This kind of exercise is more satisfactory if the regional picture is presented to begin with, Figure 5.7 being derived from U.S. Soil Conservation Service maps of Davis-Weber and Salt Lake counties, Utah. Having shown the spatial disposition of the major soil categories, topographic profiles can be con-

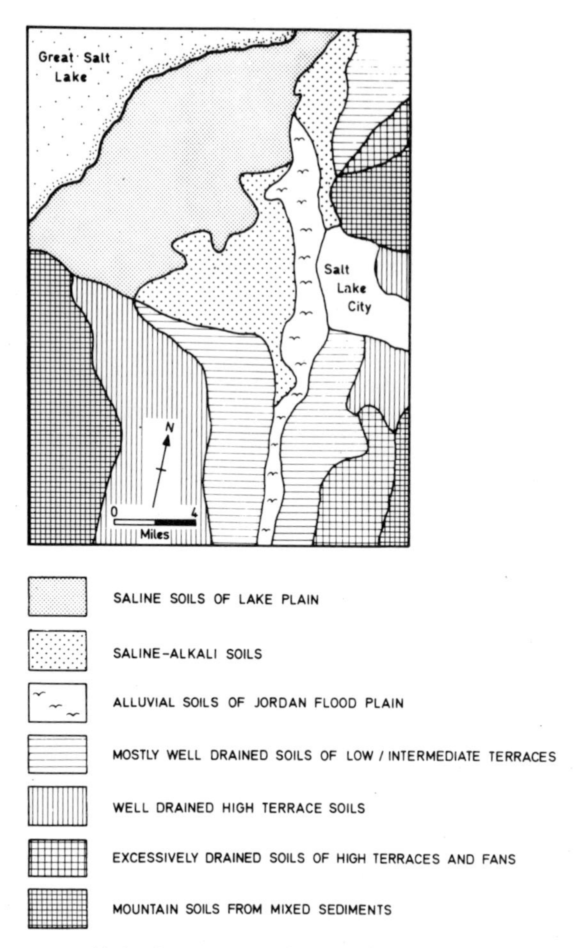

FIGURE 5.7 *Soil-landform categories, Utah, U.S.A.*

structed to bring out significant breaks of slope across the chronosequence. These in turn are annotated with soil profile details as suggested in relation to the Australian example in Section 5.3. The soil divisions in the Salt Lake Valley example are Pleistocene or Recent in origin, with the highest Lake Bonneville shoreline representing the oldest surface. The freely-drained terrace soils are predominantly of chestnut type (mollisols). Soils on the earlier terraces exhibit complete removal of soluble salts from the A and B horizons and a carbonate horizon at depth. Clay translocation appears to have followed this decalcification and led to the widespread formation of a

Bt horizon. Younger soils, particularly those on less well-drained, lower terrace sites exhibit alkali character and weaker structural development, while the soils of the lake plain itself vary between those with surface salt efflorescence (solonchaks) and those of slightly higher elevation (and older) which are strongly alkaline (solonetz). The alluvial soils have least horizon differentiation while certain of the mountain soils exhibit some faintly podzolic characteristics.

Although the climate and parent materials will have determined the rate of the soil evolutionary process, and although most of the soils have now been modified by man, the essential differences can be viewed as a function of time. While initially accessible to the student, this is undoubtedly a more advanced exercise since experienced judgement and careful consultation of the soil report are often prerequisites to identifying evidence of comparative evolution. In small localities the usual problem is that there is not sufficient age contrast between surfaces. In the selected area of Utah the soils vary greatly in age and have latterly developed under an increasingly dry climate. From this, it may be postulated that decalcification and clay movement may at one time have been occurring much more rapidly than at present or may even be 'fossil' features.

5.5 RELATIONSHIPS BETWEEN SOILS AND LAND USE

The definition of land use adopted here is broader than simply referring to use categories or land utilisation. It is extended to include details of individual crops and farming types. In Britain and many other countries, maps of land use are becoming increasingly available for exercises in soil-land use correlations. In addition, it should be pointed out that the field mapping of land utilisation of areas known to be covered by a soil map, provides a simple and interesting preliminary to analytical work in the cartographic laboratory. As with the preceding exercises, the method of overlays may be used, or, alternatively, separate diagrams can be compared.

The relationship between land utilisation and major soil groups can be explored at a general level of resolution by working on a small area with medium-scale maps. Soil series and associations need to be amalgamated into major soil groups and a tracing taken. The key of the soil map usually allows this to be done quite easily. An overlay of land utilisation categories can be abstracted for the same area.

The value of the correlation exercise at this level of generalisation lies as much in the questions it raises in the mind of the investigator as in the answers it provides. Figure 5.8 shows the major soil groups and land uses for part of Aberdeenshire, Scotland (Soil Survey of Scotland, sheet 77; Land Utilisation Survey of Britain, sheet 45). The correlation between areas of rough pasture and infertile links sand and basin peat is clearly shown, although their areas do not coincide precisely, as improvements have upgraded their margins.

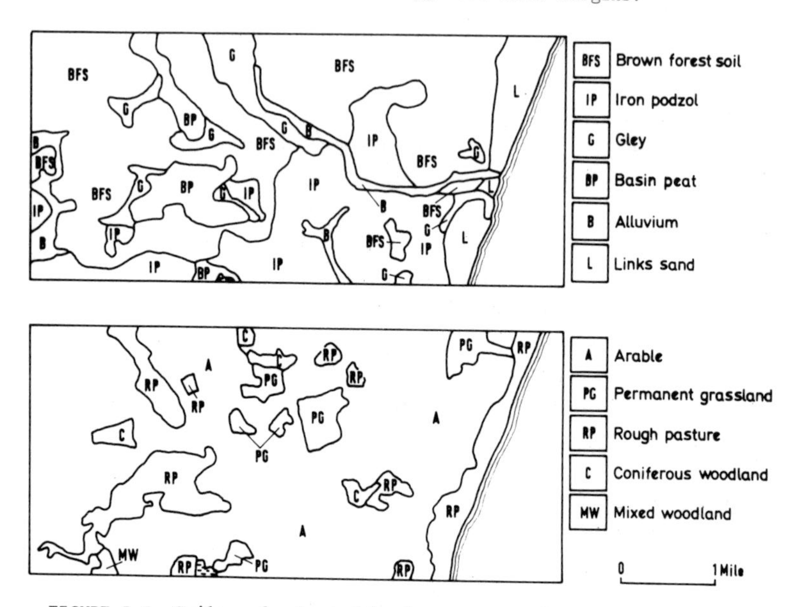

FIGURE 5.8 *Soils and selected land-use categories, Aberdeenshire, Scotland*

The major distinctions between the other soil categories - brown forest soils, podzols, gleys and alluvium - are clearly not reflected on the land utilisation map, apart from the occasional tract of rough grazing on gley soils. Elsewhere, land use categories such as arable, grassland and woodland are found distributed without any obvious regard to soil group. To a large extent this lack of correlation reflects the limitations of the scale of the maps. The soil map shows the occurrence of natural soils and does not take heed of the many management practices that can improve inherent fertility. In addition, the land utilisation map at this scale cannot give information on the many qualitative and quantitative aspects of land use that

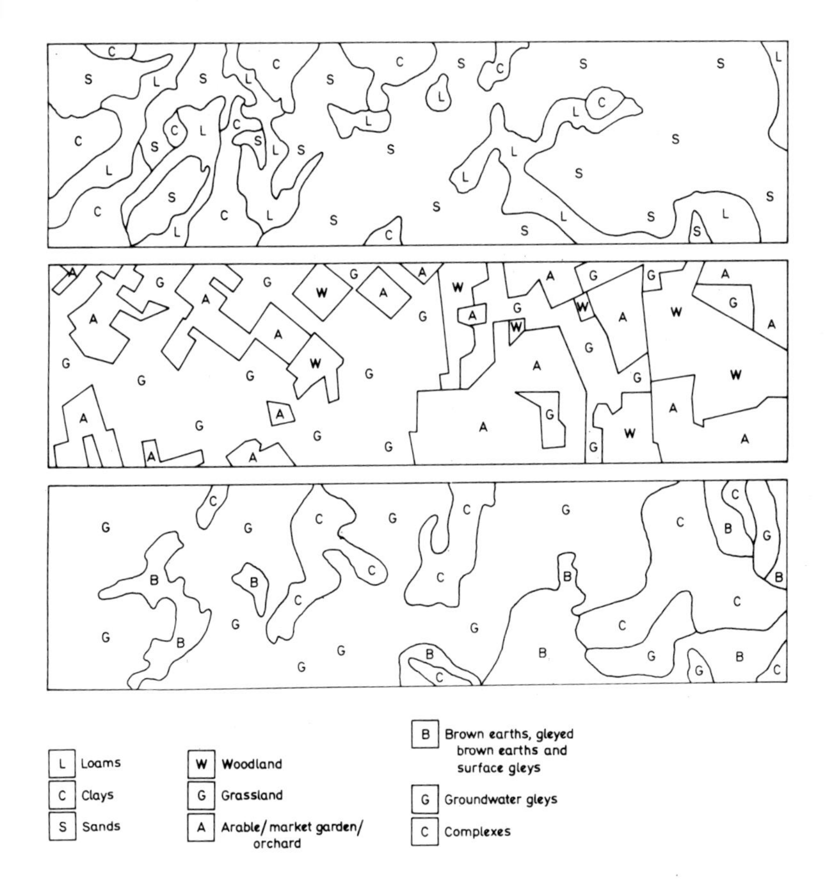

FIGURE 5.9 *Soil and land-use analysis, Vale of York, England*

might emerge from a detailed study (see also Chapter 3). For example, rotational grassland is included in the arable category, even though crops may be grown only at infrequent intervals.

At mapping scales of 1:25,000 and larger, where the soils of each field are shown, there is an opportunity to examine relationships between individual soil attributes and the way in which land is being or has been used in the past. The land use map, of course, reflects in its precise detail, economic and locational factors as well as those of an ecological kind. However, the physical limitations to farming due to soil conditions, though they may not always be reflected in differences of land use category, may nevertheless be detected in the choice of crops, type of rotation or other detailed considerations. Figure 5.9

represents separate statements on land use, surface texture and drainage for a part of the Vale of York, England (Soil Survey of England and Wales, sheet SE65; Second Land Utilisation Survey of Britain, sheet 709). An area of flat terrain lacking major relief contrasts has deliberately been selected to emphasise problems of interpretation. Pasture predominates in the west where clayey gley soils on lacustrine deposits are widespread. In the east these deposits are overlain by increasing thicknesses of aeolian or outwash sands, with woodland and arable more in evidence. The contrast is not striking, but while the whole area has a tendency to drainage impedence, the sandier textures of the east give better workability. Also many of the gley soils, particularly surface-water gleys, have responded well to drainage, so that traditional contrasts in the use of the land are fast disappearing.

The soil complexes appear to have restricted agricultural development to varying degrees. A complex is a mapping unit comprising variable soils in close juxtaposition. When isolated areas of 'complex' occur, these appear to have been improved, but the larger areas have resisted the course of agricultural improvement. In this exercise, some prior knowledge of the area will greatly improve one's perception of relationships; for instance, whether an area of woodland is in the wild state, is afforested or is managed for game, will determine whether ecological or institutional factors should be uppermost in the interpretation. The arable category is a broad one and if subdivided, shows an interesting distinction between cereal growing on the heavier soils and root crops on sands, a reflection of the fact that sugar beet would not grow so well on the heavier soils and would be more costly to lift.

Land of the highest inherent quality can sometimes be shown to have been the most intensively used or indeed the first to be cleared and improved in the course of history. The Mendip example in Figure 5.3 is an area where brown earths of the south-west facing Mendip slopes show signs of early cultivation lynchets, while many prehistoric remains are scattered across the plateau. One may speculate therefore about the effects that early forest clearance could have had on these upland loams and to what extent local podzolisation has been accelerated or even initiated by man. Much historical information can be located on detailed topographic maps, particularly of long-settled areas such as the British Isles, but recourse to primary

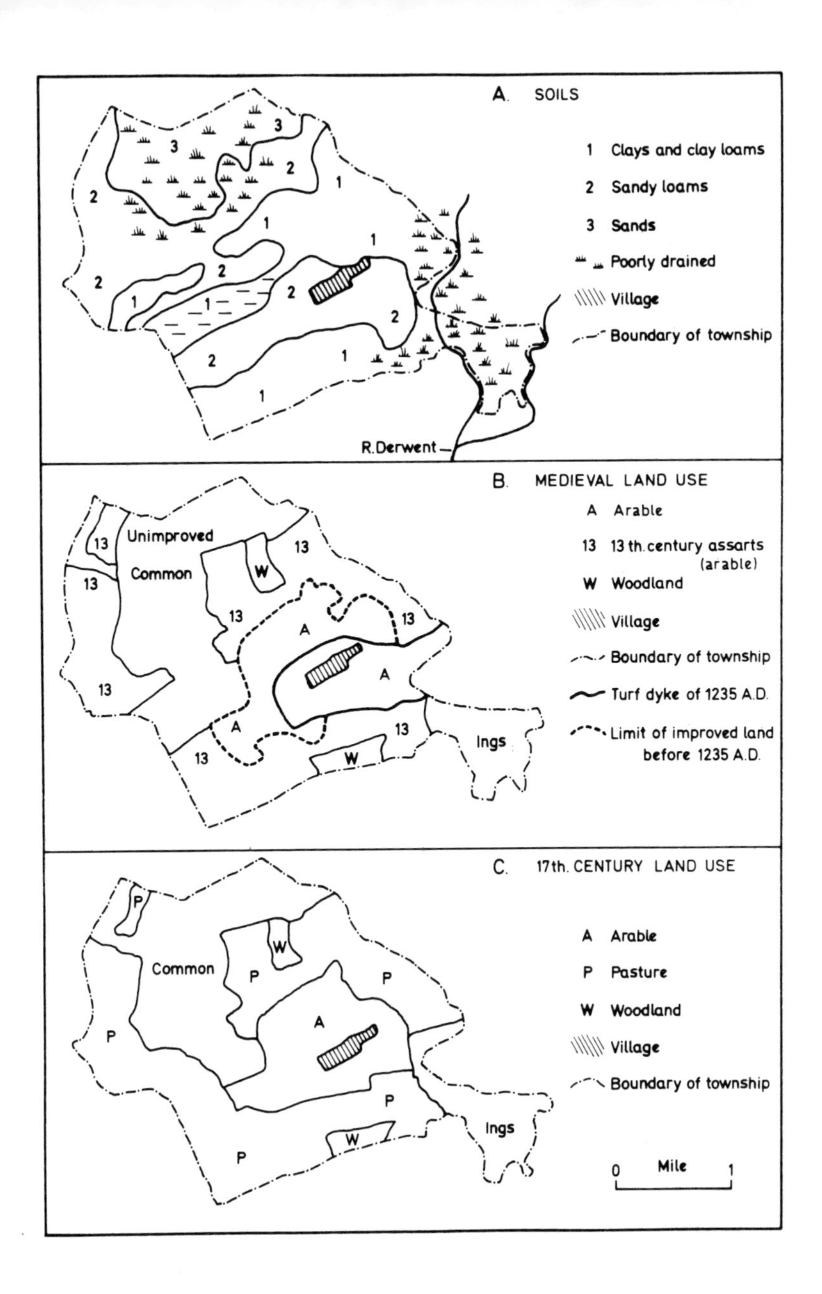

Figure 5.10 *Soils and the evolution of settlement and land-use, Vale of York, England (After Sheppard, 1966)*

historical sources is necessary for definitive local work, as the following example shows (Figure 5.10).

Wheldrake is a street village which arose as a planned settlement on the axis of the Escrick moraine, overlooking the then marshy valley of the River Derwent, to the south of York, England (Sheppard, 1966). From Figure 5.10 sandy loams (A) reflect the morainic material while a belt of heavier soils lies beyond. In early Medieval times (B) cultivation took place within the turf embankment while clearances of the forest were taking place on suitably drained areas beyond this. In the later period (C) the cultivated open field strips were still located close to the village on the sandy loams and better-drained clay loams. The area of wet sands to the north west persisted as a wilderness until, at a later date, effective drainage was installed. A number of detailed studies have been made (e.g. Jones 1966, 1971) which point to the importance of soil conditions in the foundation and evolution of settlements.

5.6 AREAL SAMPLING FROM SOIL MAPS

It is often necessary in working with soil maps to know approximately the areas of the different soil types. Whether the area of interest is the whole soil map, or merely a part of it, such as a defined number of parishes or a particular natural region, a knowledge of quantitative soil coverage can be useful in many ways, for example for correlations with land utilisation or even for rural planning purposes. If a planimeter is available for measuring areas, this can give a reasonably accurate result. However, the amount of labour involved in planimeter work is very large, and if an *estimate* of areas is all that is required it is far simpler to use areal sampling methods.

Sampling designs are composed of two elements, the *population* to be sampled, that is, the soil areas under investigation, and the *sampling units*, the parts selected for estimation. The sampling units can be points, traverses or small areas (quadrats), although in the interest of speed and labour the first two are preferred. The number of possible ways of choosing the sample is very large, although for most work with soil maps one of the six commonest designs can be chosen. These are illustrated in Figure 5.11.

The characteristics of each design are as follows:

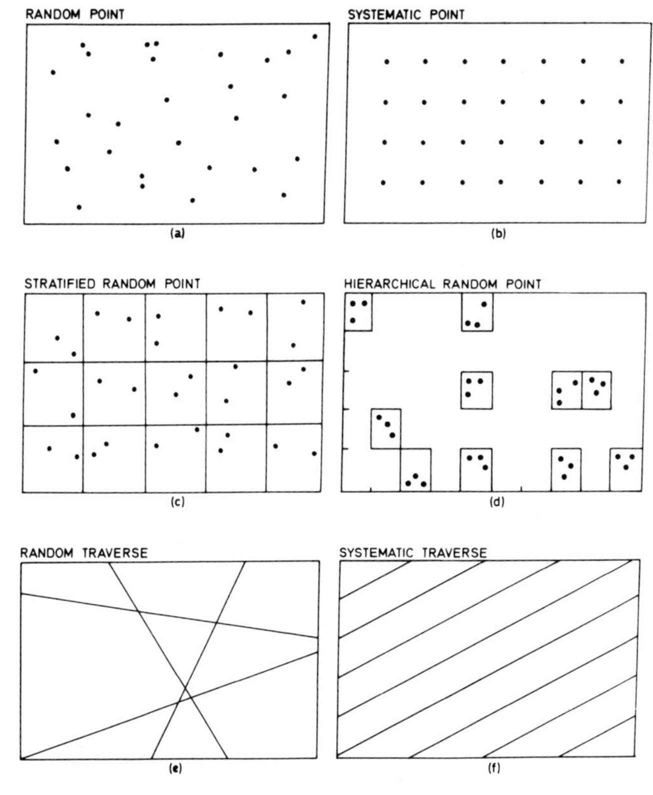

FIGURE 5.11 *Areal sampling designs*

a) Random point - all points are chosen independently of other points by using *random number tables* to give x- and y-co-ordinates for each point.

b) Systematic point - the co-ordinates of the first point are chosen at random, after which the locations of all other co-ordinates are determined by a regular interval.

c) Stratified random point - the population is divided into equal sub-areas or strata, and the points chosen at random within these strata; all of the strata contain an equal number of points.

d) Hierarchical random point (multi-stage sampling) - the area is divided into regular squares (i.e. about 50) and ten are chosen randomly; points are located randomly within the chosen squares.

e) Random traverse - traverse lines are located randomly across the area, the location of the line being fixed by using random number tables for co-ordinates.

f) Systematic traverses - the location of the first traverse is random, all others then being aligned in a regular manner.

As results of areal sampling are merely estimates of the true population, it is important to have an element of *randomness* i.e. chance in the choice of sample locations. Thus every point in the area should have an equal chance of being included within the sample. With an element of randomness it is possible to calculate the *standard error* of the sample statistic, i.e. the range within which it is known that the true value will lie with a given level of probability. The size of sample taken is also an important consideration in the accuracy of the sample statistic, as accuracy will clearly increase as sample size increases. The relationship is not a linear one, however, and there comes a point at which the labour and time involved in taking a larger sample does not give a sufficient return in accuracy to make the effort worthwhile. With point sampling, a sample size in the range of 50-100 points should be used; with traverses, approximately five traverse lines are sufficient. The standard error of areal percentage estimates is given by the expression:

$$\text{Standard error of } \%y = \pm\sqrt{\left(\frac{\%p \times \%q}{n}\right)}$$

with a 95% probability.

where n = size of sample, i.e. no. of points or traverses.

p = per cent under soil group y

q = $(100 - p)$

Standard errors for traverse sampling are difficult as there is considerable controversy among statisticians as to whether each traverse line is a sample unit (e.g. $n = 5$) or whether the sample line should be regarded as a large number of continuous points.

The standard error gives the range within which the true value will lie with a specified level of confidence. The *efficiency* of the particular design used is a measure of the *precision* of that design i.e. how closely would repeated sampling approach the true population value. The question of efficiency has been a major issue of investigation by environmental scientists and, not surprisingly, different authorities have come to favour different designs. Random traverses are not recommended, as they give no guarantee of adequate coverage.

	Random points	Systematic points	Stratified random points	Random traverse
Brown earth	32 ± 4·7	41 ± 4·9	37 ± 4·8	26 ± 19.6
Gleyed brown earth	9 ± 2·9	8 ± 2·7	5 ± 2·4	6 ± 10·6
Gley	5 ± 2·2	5 ± 2·2	5 ± 2·2	13 ± 15·0
Peaty gley	28 ± 4·5	20 ± 4·0	26 ± 4·4	13 ± 15·0
Peaty gleyed podzol	20 ± 4·0	23 ± 4·2	23 ± 4·2	42 ± 20·7
Beach deposits	5 ± 2·2	2 ± 1·4	2 ± 1·4	0
Urban	1 ± 0·9	1 ± 0·9	1 ± 0·9	0

TABLE 5.2 *Sample estimates and standard errors for soil coverage on the Aberystwyth sheet*

Table 5.2 shows the results of an exercise to find the percentage cover of different soil groups on the Aberystwyth sheet (Soil Survey of England and Wales, sheet 163). For point sampling, 100 points were taken in each case, and five traverses selected for the random traverse design. The number of sample points falling on a particular soil type is thus a direct measure of the percentage cover of that soil type in the total area, i.e.

$$\% \text{ cover of } y = \left(\frac{\text{no. of points on } y}{n} \times 100\right)$$

where n = size of sample. For traverses, lengths are measured to give

$$\% \text{ cover of } y = \left(\frac{\text{total length of traverses on } y}{\text{total length of traverses}} \times 100\right)$$

A similar criticism obtains with random points, although less severely. Systematic designs of both point and line units receive strong support from geologists and foresters on account of the regular areal coverage they both give; statisticians, however, have been less happy over the decrease in randomness of these designs, the location of all points or lines being fixed after the first one has been randomly chosen. However, for work with soil maps, both are simple, quick and sufficiently reliable. On statistical and coverage grounds they can only be surpassed by the stratified random point design, which combines randomness and efficiency, although at the cost of some extra labour. Randomness in design can be achieved by the use of *random number tables*, which are found in most collections of statistical tables. For the area to be sampled, it is appropriate to construct in pencil an x-axis (easting) and a y-axis (northing). Each of these is subdivided into 100 equal divisions. By consulting the random number tables, successive co-ordinates for the x-axis and the y-axis can be obtained by moving *vertically* down the columns of random digits.

REFERENCES

Avery, B.W. *The soils and land use of the district around Aylesbury and Hemel Hempstead*, H.M.S.O. Harpenden, 1964.

*Canada Department of Agriculture. *The system of soil classification for Canada*, Ottawa, 1970.

Duchaufour, P. *Précis de Pédologie*, Masson, Paris, 1970.

*D'Hoore, J.L. *Soils map of Africa 1:5,000,000*, Lagos, 1965.

*Food and Agriculture Organisation. *Definition of soil units for the soil map of the world*, Rome, 1968.

Franzmeier, D.P. and Whiteside, E.P. 'A chronosequence of podzols in northern Michigan', *Quarterly Bulletin Michigan State University Agricultural Experiment Station*, 46 (1963), 2-57.

*Gardiner, M.J. and Ryan, P. 'A new generalised soil map of Ireland and its land use interpretation', *Irish J. Agric. Res.*, 8 (1969), 95-109.

Glentworth, R. and Muir, J.W. *The soils round Aberdeen, Inverurie and Fraserburgh*, H.M.S.O. Edinburgh, 1963.

Haggett, P. *Locational analysis in human geography*, Edward Arnold, London, 1965.

Jones, G.R.J. 'Rural settlement in Anglesey', in Eyre, S.R. and Jones, G.R.J. *Geography as human ecology*, Edward Arnold, London, 1966, 199-230.

Jones, G.R.J. 'The multiple estate as a model framework for tracing early stages in the evolution of rural settlement' in *L'Habitat et les paysages ruraux d'Europe*, les Congrès et coloques de l'Université de Liège, 58, 1971.

King, L.J. *Statistical analysis in geography*, Prentice Hall, Englewood Cliffs, 1969.

Monkhouse F.J. and Wilkinson, H.R. *Maps and diagrams*, Methuen, London, 1964.

Muir, A. and Fraser, G.K. 'Soils and vegetation of the Bin and Clashindarroch Forests', *Transactions of the Royal Society of Edinburgh*, Vol LX(1) No. 8 (1939-40), Robert Grant and Son, Edinburgh, 1940.

Mulcahy, M.J. and Hingston, F.J. *Soils of the York-Quairading area, W.A., in relation to landscape evolution*, C.S.I.R.O. Soils Publication No. 17, Melbourne, 1961.

*Northcote, K.H. *A factual key for the recognition of Australian soils*, C.S.I.R.O., Melbourne, 1960.

Rudeforth, C.C. *Soils of north Cardiganshire*, H.M.S.O., Harpenden, 1970.

Sheppard, J.A. 'Pre-enclosure field and settlement patterns in an English township', *Geografiska Annaler*, 48B (1966), 59-77.

*United States Department of Agriculture, Soil Survey Staff, *Soil taxonomy, a basic system*, Agriculture Handbook No. 436, Washington D.C., 1975.

PART C

LABORATORY ANALYSIS OF SOIL SAMPLES

6 THE MINERAL FABRIC

6.1 INTRODUCTION

Soil material is commonly defined as a three-phase medium, consisting of solid, liquid and gaseous components. Studies of these components in the laboratory are important for two fundamental reasons. Firstly, the capacity of the soil to sustain plant growth is being increasingly recognised as a physical, as distinct from a biochemical, phenomenon. In the fields of agronomy and plant ecology, the traditional exphasis in soil studies has focused on the chemical suitability of the soil medium to provide the plant with the essential chemical nutrients that it needs. As agricultural technology has to a large extent overcome the problems of chemical infertility of soils, so emphasis has tended to shift. Physical properties related to soil aeration, soil moisture relations, erodibility and structural stability have become potentially more limiting as nutrient problems are overcome by the use of fertilisers. Details of the importance of the soil's physical characteristics in land use are given by Baver (1972) and Rose (1966).

Secondly, the arrangement of soil particles and soil aggregates in soil horizons is a reflection of pedogenic processes operating within the soil profile. As such, studies of the amount and arrangement of mineral particles leads to diagnostic interpretations of soil-forming processes which can supplement field observations. In particular, microscopic observation of the primary minerals, clay minerals and sesquioxides in soil horizons can lead to a more complete study of soil evolution than is possible from field investigation alone. Examples of the type of interpretation possible have been detailed by Brewer (1964).

The texture of soil can be accurately determined by laboratory techniques, the most widely used of which are pipette analysis (Section 6.2) and the use of a soil hydrometer (Section 6.3). Texture is important for understanding all other physical properties of soil, and thus mechanical analysis has come to be a routine laboratory determination. As plants require water for their metabolism, and as water is the medium for most biochemical processes in soils, an understanding of moisture characteristics is an essential pre-requisite for many soil studies. Water content in the field and in the air-dried state can be measured as a routine determination (Section 6.4). The estimation of the tensions at which water is held in soil over a wide range of water contents gives information on water availability to the plant as shown by the pF curve (Section 6.5). This allows prediction of the water content of soil at such critical states as saturation, field capacity and wilting point. As water is such a dynamic constituent of soil, it is important to measure the ease with which it can be transmitted through the soil body (permeability, see Section 6.6).

In order to gain insight into the behaviour of the soil under field conditions, it is desirable to conduct laboratory tests on intact soil specimens. Such tests include the estimation of density and porosity (Section 6.7) and plasticity (Section 6.8). Furthermore, indices of structural stability can be obtained by dispersion tests (Section 6.9), resistance to falling drops (Section 6.10) and aggregation after wet sieving (Section 6.11).

The microscopic examination of the mineral fabric can be carried out by three separate procedures. Under the binocular microscope the arrangement of soil peds can be readily observed (Section 6.12), thus supplementing field observation. For weathering studies, the mineral assemblage in the fine sand fraction is usually diagnostic of relations between soil and parent material (Section 6.13). Finally, the preparation of resin-embedded thin sections from undisturbed soil samples allows the study of the micro-fabrics of soil profiles (Section 6.14).

6.2 MECHANICAL ANALYSIS BY THE PIPETTE METHOD

A number of methods of particle size determination have been devised, but broadly two methods are used to determine the percentages of the different size fractions in soils. Either, aliquots can be withdrawn from a soil suspension by a pipette inserted to a specified depth (pipette method) or the decrease in density of the suspension as particles settle can be measured by a hydrometer (hydrometer method).

Both methods depend on Stokes' Law governing the settling velocities of particles in a liquid medium:

$$v = \frac{2gr^2(d_1 - d_2)}{9z}$$

where v = velocity of settling in cm/sec; g = gravity (981 cm/sec^2); r = equivalent spherical radius of particles, assuming them to be spheres; d_1 = density of particles; d_2 = density of liquid; z = viscosity of liquid.

For any given determination, g and d_2 are assumed constant. The density of particles, d_1, is assumed to be 2·65 g/cm^3. As viscosity has a large inverse effect on v, and viscosity decreases with increasing temperature, it is important to know the temperature of the suspension and keep it as uniform as possible. Thus for any determination

$$v \propto r^2 \qquad \text{or} \qquad v = kr^2$$

The sizes of mineral fractions are expressed in terms of their equivalent spherical diameters (e.s.d.), and two main schemes of grading are commonly used. The International scale (International Society of Soil Science) and the USDA scale (U.S.A. Department of Agriculture) are given in Table 6.1. The main difference between them is in the size range of the silt fraction.

Fraction		International		U.S.D.A.	
		mm	μm	mm	μm
Coarse sand	CS	2·0–0·2	2000–200	2·0–0·2	2000–200
Fine sand	FS	0·2–0·02	200–20	0·2–0·05	200–50
Silt	Z	0·02–0·002	20–2	0·05–0·002	50–2
Clay	C	< 0·002	< 2	< 0·002	< 2

TABLE 6.1 *Size grades of mineral material*

With the assumption that the density of soil particles is 2·65 g/cm^3, the settling velocities of particles in water at

$20^{o}C$ can be calculated using Stokes' Law. These calculations are presented in Table 6.2, which relates e.s.d., settling velocity, and the time taken to settle through 10 cm of water. The latter is the standard depth of sampling using the pipette.

Time taken to settle through 10 cm	Settling velocity cm/sec	Maximum diameter of particles in suspension at 10 cm depth (mm e.s.d.)
4 min 48 sec	0·0347	0·0200*
20 min	0·00833	0·0098
1 hr 15 min	0·00222	0·00506
8 hr	0·000347	0·00200†
25 hr	0·000111	0·00113

*International silt plus clay

†International clay

TABLE 6.2 *Particle size, settling time and settling velocity*

At temperatures other than $20^{o}C$, the times of settling through a depth of 10 cm will need to be adjusted. Table 6.3 shows some adjusted values, for example a time of 8 hr 0 min at $20^{o}C$ in the silt column indicates that silt-size particles settle through 10 cm in this time, and the pipette aliquot thus only contains clay-size particles.

Temperature ^{o}C	Fine sand	Silt
16	5 min 19 sec	8 hr 51 min
17	5 min 10 sec	8 hr 37 min
18	5 min 3 sec	8 hr 24 min
19	4 min 55 sec	8 hr 12 min
20	4 min 48 sec	8 hr 0 min
21	4 min 41 sec	7 hr 48 min
22	4 min 34 sec	7 hr 37 min
23	4 min 28 sec	7 hr 26 min
24	4 min 22 sec	7 hr 16 min

TABLE 6.3 *Sampling times according to temperature*

Laboratory methods of mechanical analysis involve three distinct stages. Firstly, the soil sample must be pretreated to remove all non-mineral material. Secondly, the relative proportion of the different size fractions is determined, and, thirdly, results must be calculated and expressed in suitable tabular or graphical form.

Pre-treatment of the soil sample is aimed at removing non-mineral material such as organic matter, and calcium carbonate, and at dispersing the mineral particles so that in suspension they are in a non-cemented, discrete state. 10 g of fine earth is weighed into a tall 600 or 800 ml beaker. 50 ml of hydrogen peroxide (30 vol.) is added and the mixture allowed to stand for one hour for oxidation to proceed in the cold. After that time, a further 50 ml is added and warmed gently until the reaction subsides. Excessive frothing can be controlled by one drop of capryl or amyl alcohol. If necessary, further peroxide can be added until all the organic material is decomposed. The mixture is now boiled gently for a few minutes, allowed to cool, and diluted to about 200 ml with distilled water. The suspension can now be transferred to the metal cup of a mechanical stirrer (see also Sections 7.11 and 7.12 for removal of organic matter).

If the soil is calcareous, the carbonate must be removed before proceeding to dispersion and fractionation. 25 ml of 2*N* hydrochloric acid are added to the suspension in the beaker, and the soil allowed to stand for one hour with frequent stirring. The solution is then tested to make sure that it is acid to litmus and is then filtered through a Buchner funnel (Whatman No. 50 paper) on a suction flask. The soil on the filter paper is washed thoroughly with warm water until it is free of acid. When the soil has been sucked as dry as possible, it is transferred to the metal cup of a mechanical stirrer and 200 ml of distilled water added.

For dispersing the soil, use is made of the principle of cation exchange and all clay particles are saturated with sodium ion. 2 ml of 2*N* sodium hydroxide are added to the suspension, which is then stirred for 15 minutes. The suspension is then transferred to a one litre measuring cylinder and made up to the mark with distilled water. The temperature of the liquid is taken and the time of settling required before sampling at a depth of 10 cm determined. If a special sampling stand and Andreason pipette is not available, a piece of card can be fixed to the stem of a 20 ml pipette so that when it rests vertically on the top of the cylinder, the tip of the pipette is exactly 10 cm below the surface of the liquid.

The cylinder is stoppered and shaken end over end for 1 minute, making sure that all sediment is re-suspended. The cylinder is stood on the bench and the pipette inserted 30 seconds before the time for sampling silt and clay at the appropriate temperature. The 20 ml aliquot is sucked smoothly from the suspension and released into an evaporating basin previously oven-dried and weighed. The aliquot is evaporated to dryness in an oven at 105°C, cooled in a desiccator and weighed.

$$\text{Wt. of silt plus clay in aliquot} = A \text{ g}$$

$$\text{Wt. of silt plus clay in sample} = \left(A \times \frac{1000}{20}\right) \text{ g}$$

The cylinder is now reshaken by hand and allowed to stand for 8 hours, or any other convenient time, in a position of uniform temperature. The sampling procedure is repeated after the appropriate time and the weight of clay in the sample calculated as before.

$$\text{Wt. of clay in aliquot} = B \text{ g}$$

$$\text{Wt. of clay in sample} = \left(B \times \frac{1000}{20}\right) \text{ g}$$

To determine the content of fine sand and coarse sand, the supernatant liquid is decanted and discarded, and the sediment washed into a tall-form 800 ml beaker. A mark is made 10 cm from the bottom of the beaker on the outside, and the beaker filled to the mark with distilled water. The suspension is stirred and allowed to stand for the time appropriate, according to temperature, for all the sand to settle on the bottom of the beaker (Table 6.3). The turbid supernatant liquid is poured off, care being taken not to disturb the sand sediment.

Category	A Upper limit (mm)	B % soil sample	C % mineral material	D Summation % mineral material
Coarse sand	2·0	30·5	32·6	100·0
Fine sand	0·2	22·4	23·9	67·4
Silt	0·02	15·9	17·0	43·5
Clay	0·002	24·8	26·5	26·5
Loss on ignition		5·2		
Calcium carbonate				
Difference		1·2		
Total		100·0	100·0	

TABLE 6.4 *Typical mechanical analysis results*

The processes of stirring, sedimentation and decantation are repeated until the liquid poured off is clear at the appropriate time interval. The coarse and fine sand residue is transferred to a weighed evaporating dish, dried in the oven at 105°, cooled in a desiccator and weighed. The coarse sand is determined by sieving through a weighed 0·2 mm sieve, and reweighing the sieve plus coarse sand. The fine sand fraction is carefully retained for mineralogical analysis (Section 6.13).

The results of mechanical analysis can be conveniently expressed in three different ways, either in a table, as a summation curve or on a triangular textural diagram. Table 6.4 shows a typical set of results. The results are expressed as % weight in the soil sample (column B), as % weight of mineral material (column C) and as summation % of mineral material (column D). Figures for loss on ignition and calcium carbonate can be added if available (Sections 7.11 and 7.3 respectively). The 'difference' represents the sum of experimental errors. The summation curve can be constructed from columns A and D (Figure 6.1). Finally, the textural class can be designated by reference to a triangular co-ordinate chart. Two types are illustrated in Figure 6.2, one for International and one for U.S.D.A. limits. In the present example the soil would be designated clay loam.

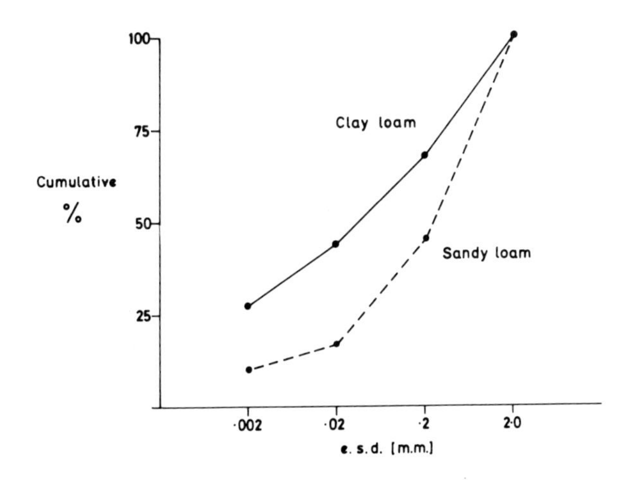

FIGURE 6.1 *Examples of summation curves*

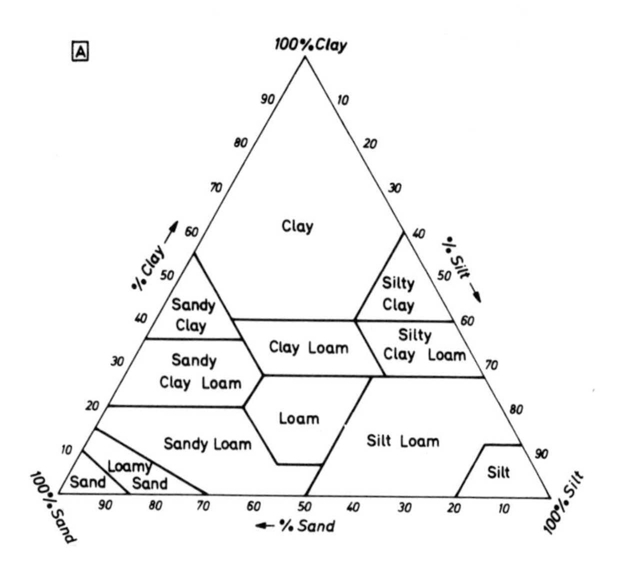

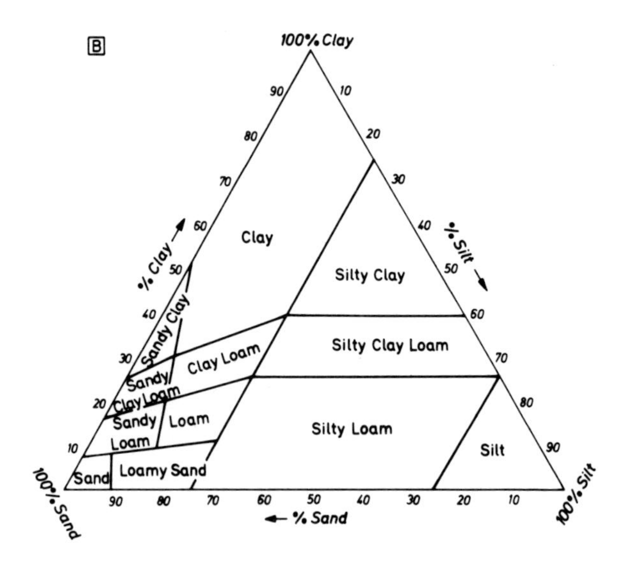

FIGURE 6.2 *Textural Triangles: United States Department of Agriculture* (A) *and International Society of Soil Science* (B)

If further data is needed on coarse and fine sand fractions for pedological or geomorphological studies, it is possible to carry out dry sieving on these two fractions for additional detail. Required sieving and weighing can be performed using chosen sieves from a range of aperture sizes (Table 6.5).

USDA SAND GRADES	Metal British Standards 410 sieves (8 inch diameter)	
	Aperture diameter (mm)	Mesh No.
Very coarse	2·0	10
1·0		
Coarse	0·6	25
0·5		
Medium	0·4	36
	0·28	52
0·25		
Fine	0·22	72
	0·15	100
	0·10	150
0·1		
Very fine	0·085	170
	0·07	200
	0·06	240
	0·05	300
0·05		

TABLE 6.5 *Sieve sizes*

6.3 MECHANICAL ANALYSIS BY THE HYDROMETER METHOD

This method was introduced by Bouyoucos (1934) and remains the basic routine method for the determination of particle sizes in soils. When used by soil laboratories for the rapid determination of texture, it is usually calibrated against the results of pipette analysis. Assumptions in the method are discussed by Black (1951).

About 50 g of fine earth are weighed into an 800 ml beaker, and 60 ml of 30 vol. hydrogen peroxide carefully added. The beaker is gently warmed until all frothing has ceased, and then gently boiled for a few minutes to destroy excess peroxide solution. Excessive frothing during the hydrogen peroxide treatment can be controlled by adding a few drops of capryl or amyl alcohol.

When cool, 10 ml Calgon (50 g Na-hexametaphosphate + 5·724 g Na_2CO_3 in 1000 ml H_2O) is added, the suspension washed into the cup of a mechanical stirrer and stirred for 15 minutes. The soil suspension is then washed into a one litre measuring cylinder and diluted to the mark with distilled water. Any remaining froth on the liquid can be dampened by one drop of capryl or amyl alcohol. The temperature of the suspension is taken, and the cylinder shaken end over end for about 1 minute, the end being stoppered by a rubber bung or the palm of the hand.

The cylinder is now placed on the bench, a stop clock started, and the hydrometer gently inserted so that it is in a position for reading. The first reading is taken after 40 seconds, and this gives % silt and clay (U.S.D.A. limits). The second reading is taken after 4 minutes and gives % silt and clay (International limits). After this stage the hydrometer is withdrawn, the cylinder shaken again and allowed to stand for 2 hours. If a water bath is available, the cylinder can be stood in this at 20°C. Just before 2 hours the hydrometer is re-introduced and a third reading taken. This gives % clay (U.S.D.A. and International limits).

The supernatant suspension is now discarded and the sediment carefully transferred to an 800 ml beaker. The sand fractions can now be studied as in Section 6.2.

The calculation of results from the hydrometer method depends partly on the calibration of the hydrometer. Some soil hydrometers are graduated in g/litre, and hence the weight of particular fractions can be read directly. Others are graduated in g/ml, giving a density reading (R_H). The density can be converted to g/litre by reading the decimals only and placing a decimal point between the third and fourth decimal places.

e.g. density R_H 1·0255 = 25·5 g/litre

It is helpful to record the calculations in a table, as shown in Table 6.6

Temperature	Elapsed time	N g/l	$N - t$ g/l	$N - t - d$ g/l	%
	40 sec				
	4 min				
	2 hours				

TABLE 6.6 *Hydrometer calculations*

The weight of mineral material in suspension (N g/litre) is firstly corrected for temperature, if necessary, as the hydrometer is calibrated at 20°C. This is done by adding 0·3 units per 1°C above 20°C or by subtracting 0·3 units per 1°C below 20°C. This gives $N - t$ g/litre. The second correction is for the density of the Calgon, and 0·5 units is subtracted from the temperature-corrected reading to give $N - t - d$ g/litre. The percentages of coarse sand, fine sand, silt and clay can now be calculated from the precise weight of soil taken. These will not add up to 100% due to the loss of organic matter, determined separately by loss on ignition (Section 7.11). Corrected percentages can be obtained by using the following conversion formula:

$$\text{Corrected \%} = \text{determined \%} \times \frac{100}{100 - \text{\% loss on ignition}}$$

The texture of the soil can now be designated by reference to the triangular diagram.

6.4 DETERMINATION OF FIELD MOISTURE CONTENT AND AIR-DRY MOISTURE

Both of these determinations are carried out in the same manner by gravimetric analysis, the former on the field sample as soon as possible after it has been brought back to the laboratory, and the latter on air-dry fine earth.

A weighing bottle or evaporating basin is dried overnight in an oven, cooled in a desiccator and accurately weighed. About 10 g of sample is accurately weighed into the container, and dried to constant weight, generally at least 6 hours, at 105°C. It is then cooled in a desiccator and reweighed.

The relevant moisture content is given by:

$$\text{Moisture \%} = \frac{\text{Loss in soil weight}}{\text{Weight of oven dry soil}} \times 100$$

6.5 MEASUREMENT OF SOIL WATER POTENTIAL USING THE PRESSURE MEMBRANE APPARATUS

Soil water potential, rather than soil water content, determines many properties of soils such as engineering behaviour at different water contents and the availability of soil moisture to plants. Each type of soil has its own characteristic *water potential curve*, commonly called the *pF curve*, which relates moisture content (in percent) to the force (in pF units) at which moisture is held in the soil. The experimental procedure is thus aimed at subjecting the soil to forces of known strength and then determining the moisture content of the sample. The determination of the entire pF curve, from approximately pF 0·3 to pF 7·0, is an elaborate procedure requiring different methods for different parts of the curve (Marshall, 1959). However, the pressure membrane is a relatively versatile apparatus covering the range from pF 3.0 to pF 4.2 (1-15 atmospheres).

The pressure membrane apparatus was first introduced by Richards (1947). A cellulose membrane, which is permeable to water but not to air, is supported on a porous base inside a steel container designed to withstand pressures over 15 atmospheres. Samples of saturated soil paste in small plastic or rubber rings are placed inside the container on the membrane, and pressure from a compressor is then applied at the required level until no more water is forced out of the samples (Figure 6.3).

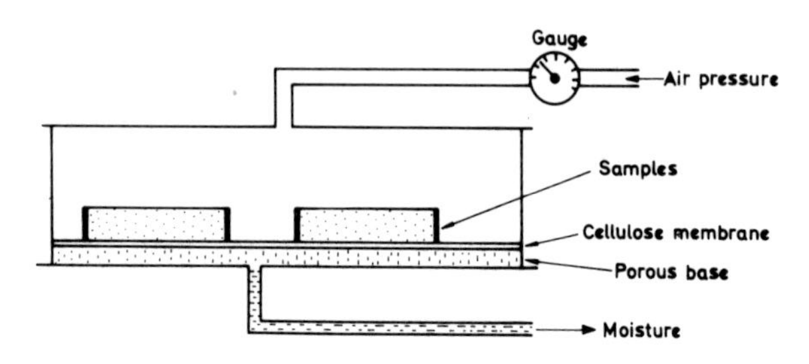

FIGURE 6.3 *Pressure membrane apparatus*

50 g of air-dry soil are placed in a basin and water added until a saturated paste is formed. It is stirred thoroughly and left to stand with the basin covered for half an hour. The cellulose membrane is soaked in distilled water and placed on the porous base of the pressure membrane apparatus. Duplicate samples of soil paste are placed in labelled rubber or plastic rings, and firmly pressed onto the membrane. The cover of the container is bolted down and the pressure regulated to 15 atmospheres.

The samples are kept under pressure for two days, before the pressure is released and the cover unbolted. The soil is removed from the container, accurately weighed, dried at 105°C in the oven, and reweighed to determine the moisture content. The experiment is repeated at 3 and 10 atmospheres pressure.

The results are best expressed in the form of a graph which relates pressure on the y-axis to moisture content % on the x-axis (Figure 6.4). Pressure can be expressed in millibars, atmospheres or the more usual pF units. Table 6.7 summarises the conversions between these. pF 4.2 is an approximate estimate of the *wilting point*, i.e. the water content when plants wilt and fail to recover. It should be noted that the moisture curve in this experiment has been determined by drying the sample under pressure; it is therefore referred to as the 'drying curve'; due to hysteresis effects, the 'wetting curve' would, if determined, lie to the left of this.

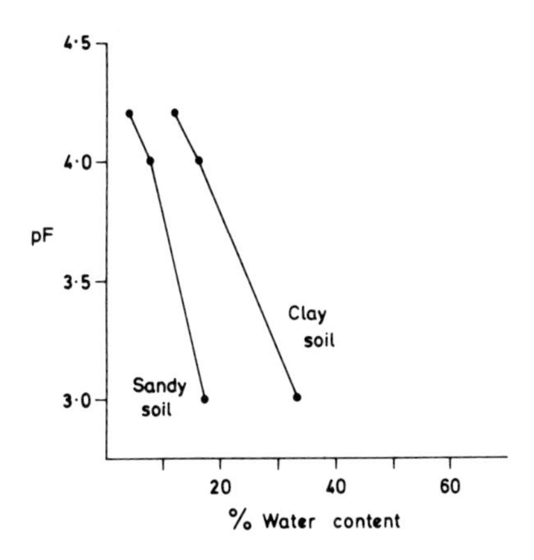

FIGURE 6.4 *Curves of soil moisture tension*

Pressure		pF
Length, h, in cm. of column of water or millibars	Atmospheres or bars	log h
1	0·001	0
100	0·1	2·0
3000	3	3·5
10000	10	4·0
15000	15	4·2

TABLE 6.7 *pF conversions*

Another important moisture constant is *field capacity*, which is defined as the amount of moisture retained by a saturated soil when free drainage has ceased. As its pF valve is low (approximately 2·5) its determination does not require the use of the pressure membrane. It can be measured by saturating, draining and re-weighing a known weight of soil.

6.6 ASSESSMENT OF SOIL PERMEABILITY

The permeability of a porous medium like soil is defined in terms of the volume of fluid or gas flowing through the medium per unit cross-sectional area, per unit hydraulic gradient, per unit time. In the following experiment the porosity, i.e. percent pore space, and permeability of a sample of sand are determined by measuring the rate at which water flows through a column of the sand in a glass tube. After measurements have been made on the sand, the experiment is repeated with soil, and the results compared.

A glass tube is taken, about three quarters of an inch in diameter and twelve inches in length. The diameter of the tube (d) is accurately measured. A piece of muslin gauze is fixed firmly over one end so that the sample will not be washed out of the tube. The tube is now filled with sand which is added slowly and the side of the tube frequently tapped to ensure uniform packing. In order to measure lengths along the tube, a paper scale is attached to it. A stand and clamps are now prepared so that the burette filled with water can be fixed above the tube with the tip about one millimetre above the sand surface; the tube itself is positioned vertically over a measuring cylinder which is standing on the bench.

The time is noted and the burette turned on. The flow of water from the burette is adjusted as rapidly as possible to provide a constant head of about one millimetre of water above the sand. The tap may require readjustment as the water level in the burette descends. A series of simultaneous readings are taken at intervals, and records made of the time (t), the distance from the surface of the sand to the wetting front (x), and the volume of water delivered by the burette (v). The data is best recorded in a table:

t sec								
x cm								
v ml								

When the water reaches the base of the sand column, it will percolate through the muslin and be collected in the measuring cylinder. A second series of readings can be taken of time against the amount of percolate collected (v^1)

t								
v^1 ml								

All readings for the measurement of permeability and pore space have now been made. First of all, the cross-sectional area of the tube (A) is calculated:

$$A = \left(\frac{\pi}{4} d^2\right) \text{ cm}^2$$

From the first series of readings, it is possible to construct two graphs, the first of x against t and the second of x against v. The first graph indicates to what extent the rate of water entry is uniform. The second graph shows the relationship between the length (x) of sand column wetted when a volume of water (v) has been added, filling the pores with water. If, in the time interval t_1 to t_2, the length of the wetted column increases from x_1 to x_2 when the volume of water added increases from v_1 to v_2, an average value $(v_2 - v_1)/(x_2 - x_1)$ can be obtained from the slope of the graph. The percent pore space is thus estimated by the following formula:

$$\% \text{ pore space} = \frac{v_2 - v_1}{A(x_2 - x_1)} \times 100$$

From the second series of readings, a third graph is plotted showing amount of percolate (v^1) against time (t). The slope of the graph can again be used to obtain an average value of $(v_2^1 - v_1^1)/(t_2 - t_1)$.

Thus the permeability of the sand for a hydraulic gradient of 1 cm water per cm is given by:

$$\text{Permeability } K = \frac{v_2 - v_1}{A(t_2 - t_1)}$$

where t is measured in seconds.

6.7 ESTIMATION OF APPARENT DENSITY, TRUE DENSITY AND PORE SPACE

Although accurate measures of the soil's constitution can only be determined in the field, reasonable estimates of apparent density, true density and pore space can be made fairly rapidly in the laboratory. The chief value of these indices is in the comparison of different soils, and hence it is usually instructive to have a range of samples to work with at the same time. The field samples are oven-dried at 105°C and passed through a

2·0 mm sieve. The crushing and sieving process should be reasonably consistent from one sample to another.

The apparent density, i.e. the density of the soil in its natural state, is determined by the use of a specific gravity bottle. The bottle is dried, cooled in a desiccator and weighed. It is then filled with the soil and reweighed. The bottle is emptied of all traces of soil, filled with water and reweighed. The apparent density may now be calculated as follows:

Wt. of soil = Wt. of S.G. bottle + soil - Wt. of empty S.G. bottle

$$\text{Volume of soil} = \left(\frac{\text{Wt. of water required to fill S.G. bottle}}{1\ \text{(density of water)}}\right) \text{cm}^3$$

where

Wt. of water = (Wt. of S.G. bottle filled with water - Wt. of empty S.G. bottle) g

$$\text{Apparent density (Bulk density)} = \left(\frac{\text{Wt. of soil, g}}{\text{Volume of soil, cm}^3}\right) \text{g/cm}^3$$

The true density, i.e. the density of soil mineral particles, is often assumed to be 2·65 g/cm^3, as in mechanical analysis, but slight variations can occur from soil to soil due to variations in the mineralogy of soil fractions. A specific gravity bottle is filled with water and weighed. 2·0 g of soil are weighed into a small beaker and the air is expelled by gently boiling with a minimum amount of distilled water. The soil is now washed into the specific gravity bottle, which is finally topped up with distilled water. The bottle and its contents are now reweighed. The calculation of true density is as follows:

Let x = Wt. of S.G. bottle filled with water

Let y = Wt. of S.G. bottle filled with water and soil

Assume 1 cm^3 water = 1·0 g

then

$$\text{True density} = \left(\frac{2}{(x + 2) - y}\right) \text{g/cm}^3$$

The pore space can be calculated using the values for apparent density (d_a) and true density (d_t). For a given volume of soil V, its weight will be Vd_a g.

$$\text{The true volume of soil particles} = \left(\frac{Vd_a}{d_t}\right) \text{cm}^3$$

$$\text{Therefore, volume of pore space} = V - \left(\frac{Vd_a}{d_t}\right) \text{cm}^3$$

The percentage pore space can now be calculated:

$$\% \text{ pore space} = V - \left(\frac{Vd_a}{d_t}\right) \times \frac{100}{V}$$

$$= V - \left(\frac{d_a}{d_t}\right) \times 100, \text{ for unit volume of 1 cc.}$$

$$\text{Therefore, \% pore space} = \left(\frac{d_t - d_a}{d_t}\right) \times 100$$

6.8 DETERMINATION OF SOIL PLASTICITY (ATTERBERG LIMITS)

Plasticity of soils refers to their behaviour, when saturated, under the stress of an applied force. The factors involved in the behaviour of a wet soil in the field are complex, but the *Atterberg limits* give a reasonable estimate of these factors by defining the range of moisture contents within which the soil is plastic. This is done by determining the *upper plastic limit* (or *liquid limit*) and the *lower plastic limit*. The former is the moisture content at which a soil paste will flow under the pressure of a small force; the latter is the minimum moisture content at which the sample shows plastic behaviour. The difference between the upper plastic limit and the lower plastic limit is referred to as the *plasticity index* (or *plastic range*).

The plastic limits can both be determined on the same soil paste. About 50 g of dry soil are spread out on a flat plate and water slowly added. The mixture is stirred with a spatula and, when it approaches the plastic state,is kneaded between the fingers after each addition of water, so that the moisture content is homogeneous. Water is added until a stiff paste is formed.

The upper plastic limit is determined by the Casagrande liquid limit apparatus (British Standard, 1967). This is designed to reduce subjectivity in the determination. A brass pan is connected to a cam which raises it to a height of one centimetre and then allows it to drop onto a hardened rubber surface as the crank is turned. The pan is filled with paste to a depth of 1 cm and levelled off. The grooving tool is drawn across the surface along the diameter, thus producing a gap 2 mm wide in the paste. The crank is now turned at a rate of two rotations per second and the number of shocks counted, before the two sectors of paste come into contact along a minimum distance of 1·5 cm. A sample of paste is now used to determine its moisture content by weighing, oven-drying and then reweighing. The procedure is now repeated at least four times, using different moisture contents of the paste in order to give a range of shocks from about 50 to 10 to produce flows. A graph can now be drawn showing the number of shocks against moisture content, so that, by interpolation, the moisture content for flow after 25 shocks can be determined. This, by definition, is the upper plastic limit.

The lower plastic limit is determined by spraying water onto a piece of the original soil paste until it can be rolled into a ball in the palm of the hand. It is then rolled between the hand and a piece of glass to form a thread. When the diameter of the thread is 3 mm, it is rekneaded and rolled again. This process is continued until the thread crumbles when the diameter is 3 mm. The moisture content when this occurs is the lower plastic limit. It is customary to repeat the experiment at least once and obtain the mean value.

6.9 DETERMINATION OF THE DISPERSION RATIO

The dispersion ratio gives a rapid assessment of the stability of soil structural units when subjected to severe shaking in water. It therefore represents a measure of the ability of soil structural units to withstand severe waterlogging under field conditions.

10 g of an air-dry field sample is placed in a measuring cylinder and distilled water added to make the volume up to one litre. The cylinder is stoppered and thoroughly shaken end over end for one minute. The cylinder is placed in a water bath at 20°C and the suspension allowed to settle for 40 seconds, i.e. until a 20 ml sample pipetted from a 10 cm depth consists of particles with a maximum size of 0·05 mm e.s.d. The pipetted fraction is placed in a weighed evaporating basin, evaporated to dryness, weighed, and the weight of silt plus clay calculated. From this, the % dispersed silt plus clay can be calculated.

The dispersion ratio is the ratio of this weight to the total silt plus clay content as determined by mechanical analysis.

$$\text{Dispersion ratio} = \frac{\%\ \text{dispersed silt + clay}}{\%\ \text{total silt + clay}}$$

The value of this technique is largely as a comparative measure of different soil types. Estimations on a range of soils can bring out important differences in structural stability.

6.10 MEASUREMENT OF SOIL AGGREGATE STABILITY BY THE IMPACT OF FALLING WATER DROPS

The breakdown of soil aggregates under the impact of falling water drops is an important process under field conditions, and can be conveniently simulated in the laboratory to give an index of structural stability. Low (1954) has calculated that raindrops of average size 4 mm diameter have a terminal velocity of 7 m per second, thus dissipating $8{\cdot}2 \times 10^3$ ergs per drop on impact. In the following experiment, waterdrops of weight 0·1 g are allowed to fall one metre onto moist aggregates, giving a kinetic energy on impact of $9{\cdot}3 \times 10^3$ ergs.

A 50 ml burette is set up so that a constant level of water can be maintained in it by an overflow fitted 5 cm from the top. The tap of the burette is fitted with a rubber jet, of diameter 4·25 mm. Air-dry soil aggregates between 4 and 5 mm diameter are separated from the field sample by sieving and placed on filter papers resting on blotting paper covering a dish of coarse sand. The dish of coarse sand is wetted until the water table is about 5 cm from the sand surface. The aggregates are allowed to stay on the filter paper until they are of constant weight. This method of wetting the aggregates is specifically designed to moisten the aggregates without disrupting their structure in any way (Low, 1967).

When moistened, the aggregates are placed on a wire gauze with holes approximately 3 mm diameter. The gauze is placed over a beaker and arranged so that it is one metre below the tip of the burette. Water drops are allowed to fall at the rate of 50-70 per minute, and the number required for breaking down the aggregate so that it is washed through the 3 mm hole recorded. At least 10 determinations should be made on each soil sample and the average values compared for different soils.

6.11 DETERMINATION OF WATER STABILITY OF AGGREGATES BY WET SIEVING

The stability of soil aggregates under the relatively strong stresses of shaking in water has been one of the commonest structural properties examined in the laboratory. The technique involves simultaneously shaking and sieving a soil sample which is immersed in water; this rather violent treatment breaks down aggregates, and the degree of destruction is determined by recording the amounts of material retained by a nest of sieves (Williamson, Pringle and Coutts, 1956).

The following method has been detailed by Tinsley and Coutts

(1967), which makes use of a wet sieving machine (Figure 6.5). This apparatus contains a nest of sieves with 2 mm, 1 mm and 0·5 mm apertures. Field samples of soil are air-dried and passed through a 3 mm sieve. A sample of the aggregates retained on the 3 mm sieve is weighed and re-wetted by placing on a 0·2 mm aperture sieve whose base is just immersed in a water bath.

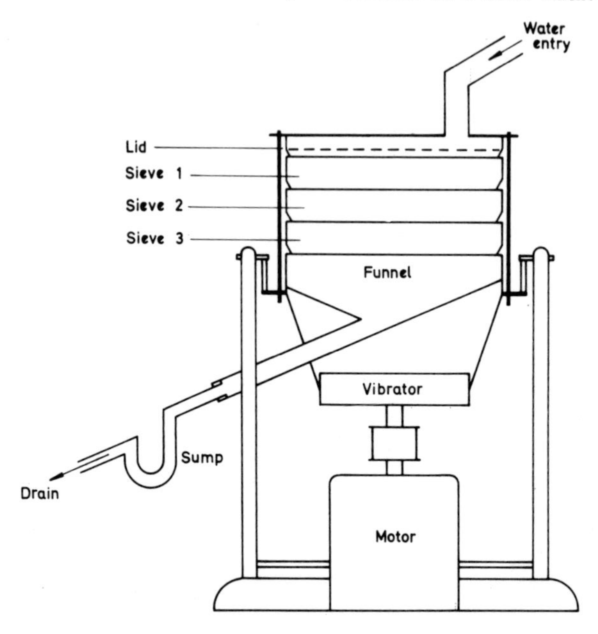

FIGURE 6.5 *Wet sieving machine (after Tinsley and Coutts, 1967)*

Samples of about 100 g are used for sands, 75 g for loams, and 50 g for clays (W).

The wetted sample is transferred to the top 2 mm sieve and the machine run for 10 minutes. Preliminary trials will have to be run with the machine so that the force of shaking can be adjusted to give complete breakdown of the sample in 10 minutes. Also, the flow of water through the machine will have to be adjusted to give a rate of flow of 4·5 litre/min.

After shaking, the three sieves are removed and the material on each, washed through a plastic funnel into separate collecting beakers which have previously been weighed. The beakers are dried in the oven, cooled and weighed, and the material retained by each sieve determined. The summation of the three fractions gives the *total structure* (S).

The aggregates in the beakers are now re-wetted and stirred into a paste. Each fraction is washed onto a weighed 0·5 mm sieve and washed through. Coarse sand grains are retained by the sieve, which is dried and weighed. The weight of primary particles (P) thus needs to be subtracted from *total structure* (S).

Results are calculated as follows:

$$\%\ \text{True crumb structure} = \left(\frac{S - P}{W}\right) \times 100$$

Four replicate determinations are recommended for each soil type, and the average values of different soils compared.

6.12 EXAMINATION OF SOIL UNDER THE STEREOSCOPIC MICROSCOPE

The examination of soils under the stereoscopic microscope is a useful exercise as a preliminary to examination by the petrological microscope (see Sections 6.13 and 6.14) and to studies of soil structure (see Sections 6.10 and 6.11). The technique is simply one of increasing the information on a particular field sample or soil fraction by viewing it under increasing magnifications under the stereo microscope. Convenient magnifications are ×5, ×15, ×30 and ×80. Notes are systematically made on the mineral and organic material, and in particular, observations recorded on the degree of intimacy between them. As with all microscope work, diagrammatic sketches help to give a systematic analysis. Although simple to perform, observation by stereo microscope should not be underestimated; it greatly increases the range of what is visible in soils and differs from many other methods of analysis where the field nature is destroyed in the determination.

6.13 MINERALOGICAL ANALYSIS OF THE FINE SAND FRACTION

The mineralogical study of the fine sand fraction (200 - 20 μm e.s.d.) is a useful technique for gaining information on the pedogenic processes operating in specific profiles. The relative frequency of mineral species can be compared between soils and between different horizons in the same soil. This gives important clues to such pedological problems as: (a) the intensity of weathering in soils, as illustrated by the frequencies of non-resistant and resistant minerals; (b) the influence of parent material on soil character, as shown by a comparison of the mineralogy of the parent material with the mineralogy of overlying soil horizons and (c) the possible occurrence of old buried soil profiles or the truncated remnants of profiles, as shown by sharp discontinuities in mineral frequencies. Details of the importance of mineralogical studies in soil science are provided by Brewer (1964).

Mineral counts are made on the fine sand fraction. In most cases, mechanical analysis will also be carried out on the sample, and thus the fine sand will be obtained in this way (Section 6.2). In cases where the soil sample is not already fractionated, the fine sand can be obtained by the usual sequence of:

1. Removal of organic matter by hydrogen peroxide
2. Dispersion by sodium hexametaphosphate (Calgon) or sodium hydroxide
3. Sedimentation and separation by sieving.

The sequence of operations carried out on the fine sample comprises specific gravity separation, mounting, identification and counting. Specific gravity separations divide the fine sand into *light minerals* and *heavy minerals*. The greatest variety of minerals occur in the heaviest fraction, and many workers restrict themselves to a *heavy mineral analysis*. However, in the interests of gaining a complete picture of the mineral content of soils, it is recommended that both light and heavy minerals are identified.

Specific gravity separation funnels are normally used for this separation (Milner, 1962), but effective separations can be made in an evaporating dish. The experiment is normally performed in a fume cupboard. Not more than 10 g of fine sand are placed in a 4-inch diameter dish and bromoform (S.G.2.84) gently poured in. The dish is covered, allowed to stand for 15 minutes, stirred, covered and allowed to stand for 30 minutes. The floating grains are carefully removed with a teaspoon and placed in a separate dish. The liquid is now filtered so that any remaining floating grains are retained on the filter paper, and the heavy residue left behind in the dish. Both fractions can now be washed in alcohol to remove bromoform, dried and weighed.

Many methods and materials are available for mounting mineral grains, but the following is convenient. One drop of 0·1% gelatin solution is placed on a clean dry slide and dried at 80°C. A solution is prepared of 10 ml distilled water, 5 ml acetone and 2 ml 2% formalin. One drop of this is added on the film, the sand grains spread evenly with a needle, and the slide dried at 80°C. One drop of Canada balsam (Refractive Index 1.54) is added and the slide heated on a hot plate. A warmed cover glass is finally pressed on.

Group	Sub-group	Mineral	Specific gravity	Refractive index
Quartz		Quartz	2·7	1·54
Felspars		Orthoclase Microcline Plagioclase	2·5 to 2·9	1·52 1·52 1·55
Ferromagnesian minerals	Olivines	Olivine	3·3	1·70
	Pyroxenes	Diopside Augite Enstatite Hypersthene	3·1 to 3·6	1·67 1·69 1·66 1·69
	Amphiboles	Hornblende	3·0	1·66
	Micas	Muscovite Biotite	2·9 2·9	1·56 1·58
Metamorphic and accessary minerals		Garnet Staurolite Apatite Sphene Zircon Rutile Fluorite Tourmaline Epidote	3·8 3·7 3·3 3.5 4·7 4·2 3·1 3·1 3·3	1·80 1·74 1·64 1·90 1·96 2·70 1·43 1·65 1·85
Opaques		Iron and mixed oxides	3·5	--
Carbonates		Calcite Dolomite Siderite	2·7 2·9 3·9	1·49 1·51 1·61

TABLE 6.8 *Common primary minerals in soils*

The identification of mineral grains can now be carried out. For those who are unfamiliar with the use of the polarising microscope and the recognition of optical properties of minerals, the guides given by Smith and Wells (1956) and Kerr (1959) will prove to be very useful. Table 6.8 lists the commonest minerals found in soils, together with their specific gravities and average refractive indices. It is usual to find that significant amounts of mineral particles cannot be identified in soils, due to corrosion and weathering; thus it is always necessary to have an 'unidentified' category for counting.

Grain counts are carried out by the line-count method. The slide is moved randomly across the microscope field and grains that touch the cross hairs are identified and counted. Research workers in soil mineralogy differ in their recommendations on the size of sample to be counted; 300 grains per sample (Milner, 1962) obviously provide less accuracy than 1200 grains per sample (Brewer, 1964). The balance between accuracy and time depends on the aim of the investigation, and 100 grains per sample may be enough for an exploratory piece of work. Frequency tables and frequency histograms are the normal methods of displaying results.

In addition to comparing mineral frequencies between soils and between horizons, the results of mineral analysis can also be used to produce *weathering indices* which attempt to assess the rate and type of weathering. A common approach is to analyse the percentage frequency of a relatively stable mineral, for example, zircon, quartz or tourmaline and use this as an index of weathering (FitzPatrick, 1971). The higher the content of a stable mineral, the more weathered is the soil. Ruhe (1956) has proposed two weathering ratios, one for the heavy minerals and one for the light minerals, with each mineral species expressed as a percentage.

$$\text{Weathering ratio for heavy minerals} = \frac{\text{zircon + tourmaline}}{\text{amphiboles + pyroxenes}}$$

$$\text{Weathering ratio for light minerals} = \frac{\text{quartz}}{\text{felspars}}$$

6.14 THE PREPARATION OF THIN SECTIONS OF SOIL

The study of soil micromorphology by means of thin sections has now become one of the major fields of practical soil science. The study of the soil *in situ* under the petrological microscope has brought a wealth of new insights into studies of soil formation and soil mineralogy (FitzPatrick, 1971) and the physical constitution of soils (Brewer, 1964). Two points are worthy of note. Firstly, the petrological microscope can be used with magnifications up to ×100 in most soil studies, thus allowing mineral identification amd micro-fabric analysis. Secondly, thin sections are made from undisturbed samples, taken and maintained in their original form and structure. Thus analysis relates directly to field conditions, a fact which gives added importance to interpretations and conclusions.

Since Kubiëna's (1938) original work using Canada balsam dissolved in xylol and kollolith, a wide range of natural resins and polymerising plastics have been used for impregnation. Natural resins do not require that samples be thoroughly dried before impregnating, but on the other hand they usually present problems in cutting, grinding and polishing, due to the fact that they are soluble in some nonpolar lubricating liquids (e.g. kerosene). The use of plastics is most popular, and generally follows the procedure of Bourbeau and Berger (1947) for Castolite. Basically, a catalyst is added to a plastic in proportions that give controlled polymerisation, and a diluent is added to reduce viscosity. Many different brands are in use, depending on the country of origin and familiarity, but the following can be recommended as a convenient and effective method.

The sample to be impregnated is dried in the oven at 105°C for 24 hours. It is then cooled in a desiccator. The plastic (Ciba Araldite CY 219) is mixed with the catalyst (Versamid 140) on a 1:1 volume basis, and thoroughly stirred for 5 minutes. The diluent (toluene) is added at a rate of 1 part mix: 3 parts toluene, and thoroughly stirred for 5 minutes. The impregnating mixture is now put under a vacuum of 1 atmosphere for 5 minutes in a vacuum desiccator or vacuum oven, so that all air bubbles are removed.

The sample is placed in an evaporating dish and covered by the impregnating mixture. The dish is placed under vacuum at 1 atmosphere and allowed to impregnate for 3 hours. The specimen can now be hardened by heat in the oven or on a hot plate at 40°C for 8 hours and then at 100°C for 1 hour.

Thin sections are prepared by cutting, grinding and polishing. A section of the specimen is cut by a diamond saw and one side polished on lapping plates using finer and finer grades of carborundum, finishing off on aluminium oxide power. This is the longest stage of the process, and it is of great advantage to have a grinding wheel on the cutting machine. Once prepared, the polished face is fixed to a glass slide by means of the original mixture (refractive index = 1·54), or by thermoplastic cements, such as Lakeside 70. Once affixed, the thin section is finally ground down and polished on lapping plates in the same way. The final desired thickness of the section should be the same as petrological slides (50 μm - 30 μm), especially if mineral identification is required, but occasionally thicker sections (100 μm) give good indications of soil fabric. The 30 μm thickness is indicated by quartz grains showing first-

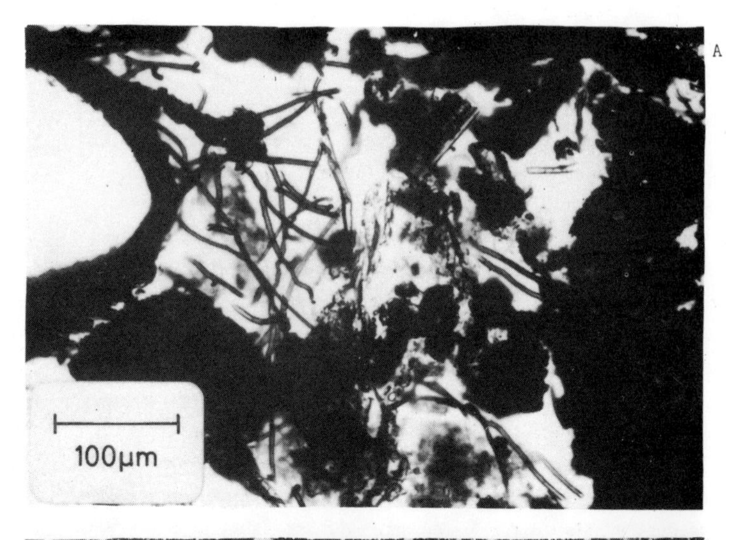

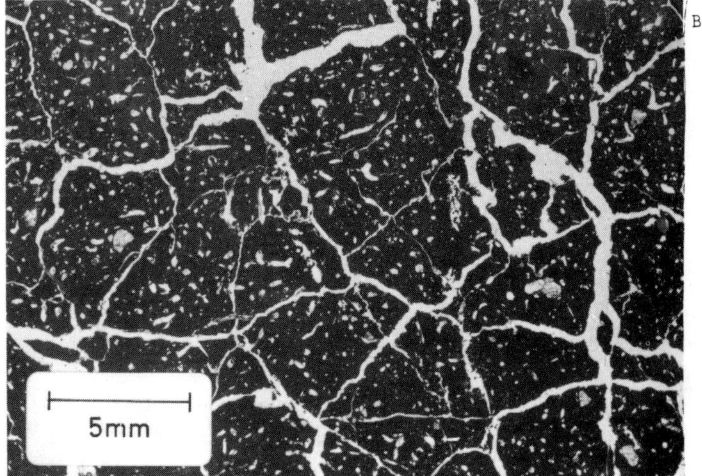

PLATE 7 *Photo-micrographs of thin sections. A. Fungal mycelium in soil pore. B. Angular blocky peds showing structure faces and internal pores*

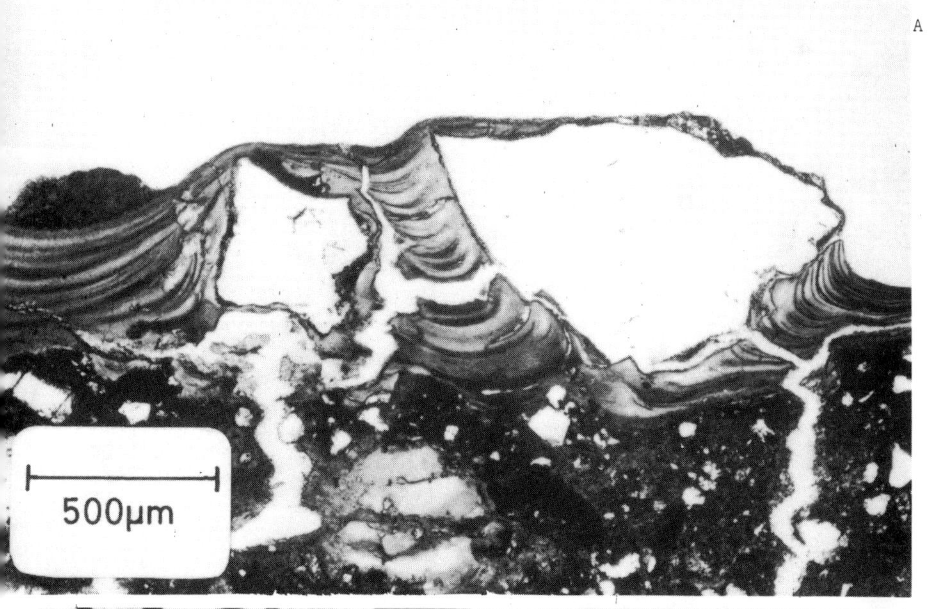

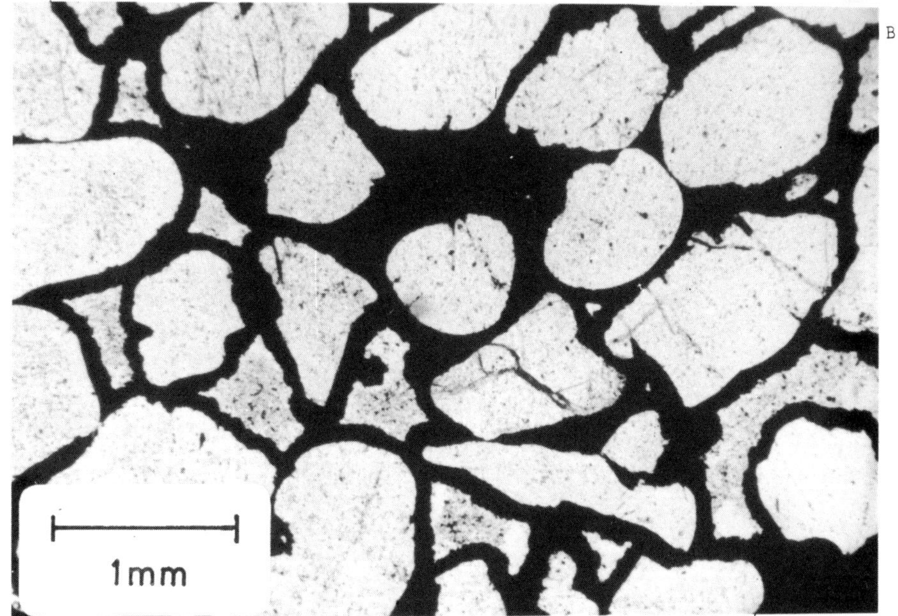

PLATE 8 *Photo-micrographs of thin sections. A. Clay skins (cutans) bridging quartz particles. B. Total cementation of quartz sand by hematite*

order interference colours of grey, white or pale yellow.

Thin sections are examined under the petrological polarising microscope both under plane polarised light and under crossed-nicols. As with most microscope work, it is advisable to make a preliminary examination at a low magnification and then to study at the usual working range of ×30 to ×50. It is important to attempt to sketch what is visible, as this helps to provide a systematic examination of soil micromorphology. By far the most comprehensive reference on thin section description and interpretation is provided by Brewer (1964). It is suggested that a systematic series of notes be made on each of the following features:

1. Soil grains, i.e. sand and silt particles
2. Voids
3. Micro-peds, i.e. aggregated units
4. Cutans, i.e. plasma concentrations such as clay skins
5. Mottles
6. Concretions
7. Roots and humic material
8. Fecal pellets

If facilities for photo-micrography are available, excellent photographs may be taken for study and display. Plates 7 and 8 show four photographs of thin sections.

REFERENCES

Baver, L.D. *Soil physics*, (4th Edn.), Wiley, New York, 1972.

Black, I.A. 'Theoretical errors of hydrometer methods for the mechanical analysis of soils', *J. Soil Sci.*, 2 (1951), 118-33.

Bourbeau, G.A. and Berger, K.C. 'Thin sections of soils and friable materials prepared by impregnation with the plastic Castolite', *Soil Sci. Soc. Am. Proc.*, 12 (1947), 409-12.

Bouyoucos, G.J. 'The hydrometer method for making mechanical analysis of soils', *Soil Sci.*, 38 (1934), 335-43.

Brewer, R. *Fabric and mineral analysis of soils,* Wiley, New York, 1964.

British Standards Institution. *Methods of testing soils for civil engineering purposes*, B.S. 1377, 1967.

FitzPatrick, E.A. *Pedology*, Oliver and Boyd, Edinburgh, 1971.

Kerr, P.F. *Optical mineralogy*, McGraw-Hill, New York, 1959.

Kubiëna, W.L. *Micropedology*, Ames, Iowa, 1938.

Low, A.J. 'The study of soil structure in the field and the laboratory', *J. Soil Sci.*, 5 (1954), 57-74.

Low, A.J. 'Measurement of the stability of moist soil aggregates to falling water drops' in *West European methods for soil structure determination*, International Society of Soil Science, Ghent, 1967, VI78-79.

Marshall, T.G. *Relations between water and soil*, Commonwealth Bureau of Soils, Technical Communication No. 50, Commonwealth Agricultural Bureaux, Farnham Royal, Bucks, 1959.

Milner, H.B. *Sedimentary petrography*, Allen and Unwin, London, 1962.

Richards, L.A. 'Pressure membrane apparatus: construction and use', *J. Agric. Engng.*, 28 (1947), 451.54.

Rose, C.W. *Agricultural physics*, Pergamon, Oxford, 1966.

Ruhe, R.V. 'Geomorphic surfaces and the nature of soils', *Soil Sci.*, 82 (1956), 441-55.

Smith H.G. and Wells, M.K. *Minerals and the microscope*, Murby, London, 1956.

Tinsley, J. and Coutts, J.R.H. 'Aggregate stability determination to determine the true crumb structure' in *Methods for soil structure determination*, Ghent, 1967, V175-77.

Williamson, W.T.H., Pringle, J. and Coutts, J.R.H. 'Rapid method for determination of water-stable aggregates in soils', *Journal of the Science of Food and Agriculture*, 7 (1956), 265-69.

7 CHEMICAL PROPERTIES AND ORGANIC MATTER

7.1 INTRODUCTION

Chemical analyses have become an integral part of pedology whether one is studying nutrients and the leaching process, pollution or more general problems of classification and genesis. But however advantageous analytical data may be, its interpretation is often far from simple. This is a problem which will be alleviated if attention is paid to some fundamental points. For instance, field variation is such that though accurately determined, the results from isolated samples may be of little value unless rigorous sampling and subsampling procedures have been followed. This admittedly is not a problem of the analysis but it should discourage the belief that the most accurate method is always the most suitable.

Soils are chemically heterogeneous media and it is generally more important to measure a particular fraction of an element in the soil than to know its total concentration. It is therefore essential to appreciate the ways in which elements may be chemically combined within the soil and to know which extractants are suitable for obtaining a specific portion of an element. For instance, in the determination of *exchangeable* cations, ammonium acetate is used in order to substitute ammonium in place of existing adsorbed cations. However, for measurement of nutrients one selects, as in the case of phosphorus, iron or potassium, an extractant which will most closely resemble the activity of plant roots under given soil conditions and thus represent what is actually *available* to the plant. For most purposes therefore, dilute solutions of weak acids are preferred. A given method may only apply to certain types of soil. In some cases, such as the determination of various salts, this is comparatively obvious while in others it may not be so straightforward. For any determination the extraction details should be recorded since the amount of an element obtained will depend on the nature of the extractant, its concentration, the ratio of soil to extractant, the manner of extraction and the length of the extraction period. It should therefore be realised that the chemical and instrumental methods of analysis available to soil scientists are very much more precise than the process of extraction. Once again, it is therefore clear that one should not be unduly preoccupied by accuracy imposed in the final stage of a technique. Where appropriate, a choice of extraction procedure has been offered in the subsequent sections.

The chemical elements in soil originate from rock minerals, rainfall or fertilisers while their concentration and combination within the soil are determined by the soil's internal climate, together with parent material, siting and biotic factors. The following techniques begin with the measurement of soil reaction (Section 7.2) as an index of soil chemistry. This test is applicable to all soils, with different factors contributing to pH in different parts of the range. Soil reaction may vary according to season or, as in acid sulphate soils, fluctuate dramatically, so it cannot be regarded as a fixed attribute of the soil. One factor governing pH is the presence of salts and the succeeding methods (Sections 7.3, 7.4 and 7.5) deal specifically with the more common examples. The presence of free salts in the soil may be due to parent material, as in the case of limestone, but over wider areas it is a function of soil climate and the influence of groundwater. The level of exchangeable bases in the soil (Section 7.6) also determines the soil's reaction and is limited itself by the cation exchange capacity (Section 7.7) of the clay and humus. Exchangeable ions are an important indication of the nutrient reserves in soils without free salts. Together with pH, the degree of saturation by individual bases can be used for studying variations in leaching between soils. This arises from the fact that the cations themselves form a series of progressively greater resistance to displacement by hydrogen ions.

Copper, lead and zinc (Section 7.8) are selected as examples of trace elements which have recently received attention as pollutants from both former and contemporary extractive and processing industries. Methods for measurement of the other trace elements are essentially similar so that the material presented can be used as a framework for further determinations. Iron (Section 7.9) is a nutrient element and in a great many soils is of diagnostic value in studies of soil genesis and classification. Particular attention must, however, be paid to extracting the appropriate iron fraction and the measurement of free iron oxides is principally applicable to acidic soils. Phosphorus (Section 7.10) is selected because of its importance as a nutrient. Owing to problems of immobilisation at high and low pH, methods for estimation of both organic and inorganically-combined phosphorus are included.

Organic matter may be present as undecomposed plant materials and as humus (Sections 7.11 and 7.12). The latter may be chemically complexed with clay minerals or exist in a largely free and therefore easily extractable form. Both may coexist in the same soil, while many soils, for instance podzols and chernozems, have a preponderance of one or other form of humus. Soil type will therefore determine the procedure used for the extraction and study of humus (Section 7.14). Total carbonaceous matter is determined by ignition (Section 7.11) while wet oxidation methods (Section 7.12) can also be used to estimate organic matter and organic carbon. The latter, however, only recover the easily digestable fraction of organic carbon. The organic matter content of soils reflects the balance between production and decay and is a function of temperature and rainfall régimes, hydrology, nutrient status and the nature of the land use and its management. The more rapidly decomposition occurs, the more efficient will be the mineralisation process and the lower the carbon:nitrogen ratio. Nitrogen measurement (Section 7.13) is of particular relevance in this context and as an index of soil fertility, although the flux of nitrogen in soil can contribute to analytical errors and demands cautious interpretation. Also, of recent importance has been the leaching of nitrate from heavily fertilised fields, leading to toxicity risks principally in inland waters.

Investigations of humus (Section 7.14) give firstly an indication of the proportion of soil organic matter which is humified, while optical studies indicate the type of humus molecules present and hence the quality of humus. This technique is of relevance in studies of intergrading soils where small changes in humus type may be significant, or simply as corroborative of other chemical data.

The method for extracting pollen and spores (Section 7.15) has been included since palynology permits the outlines of vegetational history to be established. Although originally associated with the study of anaerobic peats and sediments the method is applicable also to raw humus and acid soils. This historical component is considered essential to the full understanding of how soils and landscapes have acquired their present characteristics.

7.2 DETERMINATION OF HYDROGEN ION ACTIVITY

Hydrogen activity, soil reaction or pH is taken to be an overall indication of the chemical status of a soil. It is a statement about an aqueous solution or suspension of a soil and as such it is not to be looked upon as an intrinsic property of the soil. As an index, however, it provides a basis of comparison for all soils.

Pure water at $25^{o}C$ dissociates into the ions H^{+} and OH^{-} to the minute extent of 10^{-14}g/litre. Each ionic species contributes 10^{-7} of this ion product so that a state of equilibrium is achieved. When ionic compounds dissolve in water they generally dissociate to an extent which is determined by the substance. Those which contribute hydrogen ions are acids, and in chemical terms the strength of an acid is determined by the degree to which such dissociation occurs. The same argument applies in reverse to alkalis which donate hydroxyl ions to the solution. Materials introduced into water may therefore shift its ionic equilibrium towards acidity or alkalinity. In order that the activity or relative concentration of hydrogen ions in a solution could be expressed on a simple numerical scale without negative exponents, Sørensen devised the pH (puissance d'hydrogène) scale from 0-14 in which $pH = -\log_{10} aH^{+}$. For pure water $(-\log_{10} 10^{-7}) = 7$. Since the pH scale is logarithmic a change of one pH unit represents a tenfold change in concentration of H^{+} (Figure 7.1).

Method A - Colorimetric Two alternative procedures will be described here, both of which lend themselves to use in the field. The first method uses a testing outfit comprising glass tubes, mixed indicator solution, barium sulphate and distilled water (as supplied by B.D.H. Chemicals Ltd., Poole, England).

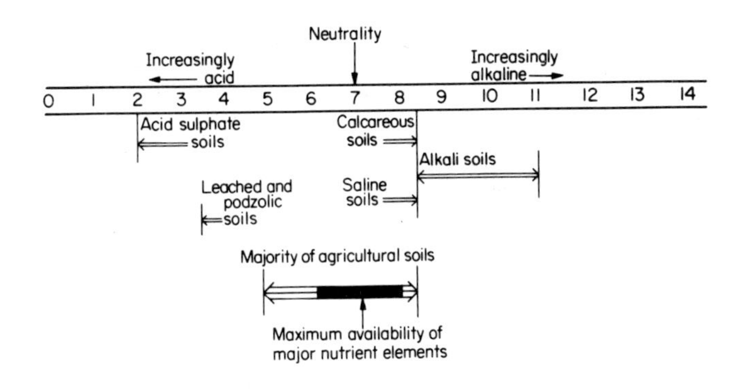

FIGURE 7.1 *The pH scale in relation to soils*

Into one of the tubes, barium sulphate powder (specially prepared for soil testing) and soil are placed in accordance with the following guidelines.

Soil type	$BaSO_4$	Soil
Sandy	1·25 cm	3·75 cm
Loam	2·5 cm	2·5 cm
Clay	3·75 cm	1·25 cm

TABLE 7.1 *Recommended proportions of soil and barium sulphate*

Distilled water is then added to the lower graduation and indicator solution to the upper graduation. The tube is stoppered and shaken vigorously, then left to stand. The colour of a clear solution which develops is compared with those of a colour chart and the pH is estimated to the nearest 0·5 pH unit. The range of pH is from 4-8. Since this is only meant to be a quick method, it is unwise to attempt interpolation between chart colours as this will give an impression of accuracy which is not justified. The judgement of colour will of course vary between individuals but it is important that the tube is held in front of a white portion of the colour chart and that direct sunlight or shadow effects are avoided. The main problem with this method is that some soils fail to yield clear solutions and others may cause excessive adsorption of the dye.

The second method uses a Hellige-Truog pH kit consisting of a white porcelain spot tile, mixed indicator solution and barium sulphate. The range is from 4-8·5 pH. For soils with pH values over 8 the La Motte pH kit is recommended (Jackson, 1958). Three drops of indicator solution are placed in one of the tile cavities and just sufficient soil is added so that the liquid is absorbed. This is mixed with a spatula and must be neither too runny nor too dry. After mixing, barium sulphate is sprinkled liberally over the soil paste so that after about one minute the indicator colour has stained the white powder. This colour is then compared with the standards printed in the booklet which accompanies the testing kit. The advantages of this method are mainly that the kit is less bulky and the procedure quicker to execute than the first method described.

Method B - Electrometric For accurate determination of pH an electrical meter is essential and in addition to the larger and more sophisticated laboratory bench models there are pH meters manufactured specifically for use in the field. These instruments are essentially millivoltmeters calibrated to read directly in pH units. The individual instrument must be set up in accordance with the manufacturer's instructions and calibrated using a pH buffer solution (e.g. pH 4, 7 or 9) which is nearest the pH range of the samples to be tested. The instrument should also be compensated for the temperature of the test solutions. This may be done manually or automated if a resistance thermometer is fitted. A nomogram for correction is supplied with some field models. The pH is measured by glass and reference electrodes which are now usually combined in one probe unit for convenience. The probe should be kept immersed in distilled water for protection and in many laboratories pH meters are kept switched 'on' so that the instrument is stabilised and ready for immediate use.

A 20 g air-dry, fine earth sample is weighed into a shaking bottle and 50 ml distilled water added, giving the standard (I.S.S.S.) soil:water ratio of 1:2·5. The bottle is stoppered and shaken mechanically for about 5 minutes or stirred regularly for about 15 minutes. The meter probe is then placed into the soil suspension and the average of three readings taken. The electrode assembly must be rinsed after each separate determination.

It is instructive to observe how pH values drift with varying dilutions, and soil:water ratios of 1:10 and 1:25 have often been used. Increasing dilution tends to shift the pH more towards neutrality. Should the glass electrode become scratched it will cease to differentiate effectively between H^+ ions and others in solution so that determinations conducted on a thin soil paste are liable to be costly.

If the pH test is conducted in 0·01M $CaCl_2$ solution instead of water (again a 1:2·5 ratio) the pH value will often be about half a unit lower than by the I.S.S.S. method. This is because $CaCl_2$ acts as a buffer solution preventing salts from ionising, and it therefore probably reflects more accurately the component of pH which derives from exchange sites on clay and humus particles.

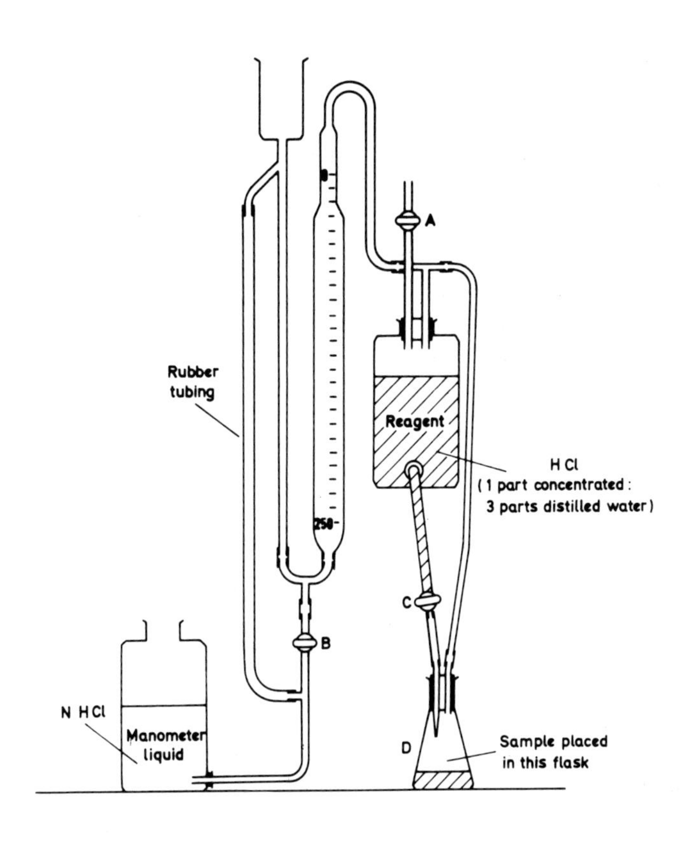

FIGURE 7.2 *The Rothamsted calcimeter (after Bascomb, 1961)*

7.3 DETERMINATION OF CALCIUM CARBONATE

Soil carbonates are sparingly soluble in water and are overwhelmingly represented by calcium carbonate or lime. For this reason such determinations are often referred to calcium unless a particular carbonate is under investigation. The first method (A) described is the titration method of Piper (1942) and the second (B) the manometric method of Bascomb (1961) to which Figure 7.2 refers. Total carbonates are measured by the amount of carbon dioxide released by the addition of hydrochloric acid. The soil is treated with an excess of HCl and in method A the excess acid is then titrated with sodium hydroxide using bromthymol blue as indicator. In method B, a measurement is made of the volume of carbon dioxide evolved.

Method A A 5 g fine earth sample is weighed into a 250 ml tallform beaker and 100 ml *N* HCl slowly added from a burette. The beaker is covered with a clock glass and stirred occasionally over a period of 1 hour. When the reaction appears to have completed and there is no further effervescence, the suspension is allowed to settle and 20 ml of the supernatent liquid is pipetted into a conical flask. The solution for titration must be as clear as possible and filtering through a fine filter paper will, if necessary, help to achieve a satisfactory result. About 10 drops of bromthymol blue indicator are added and the flask contents titrated with *N* NaOH (X ml) until one drop changes the colour to blue. The end point is quite sharp.

For the blank titration, 20 ml *N* HCl is used instead of supernatant. It is wise to perform the blank titration at least twice and obtain an average reading for the volume of *N* NaOH used (Y). As this is only an approximate method it is advisable to take an average of two to three determinations.

$$\% \ CaCO_3 = (Y - X) \times 5$$

Method B The calcimeter illustrated is more versatile and simpler to operate than the better known Collins calcimeter. It is able to take samples ten times as large (see Table 7.2), which greatly reduces the subsampling problem and there is less need for very fine soil to be used. The apparatus is manufactured privately at Harpenden, England.

Before testing the sample, the manometer liquid and acid reagent are partially saturated with CO_2 to reduce the amount of gas dissolved during determinations. This is done by decom-

Weight				CaCO3 content
20 g	for	soil	containing	5% $CaCO_3$
10 g	"	"	"	5-10% $CaCO_3$
5 g	"	"	"	10-20% $CaCO_3$
2 g	"	"	"	20-50% $CaCO_3$
1 g	"	"	"	50% $CaCO_3$

TABLE 7.2 *Recommended weights of soil for use with calcimeter*

posing $CaCO_3$ or limestone with acid in flask D and leaving for about 30 minutes with taps A and B closed.

The weight of soil to be used depends on its carbonate content. This can be roughly determined by eye or by the vigour of its reaction with HC*l*, but if in doubt, less soil is weighed, as it will take a shorter time to complete the determination.

The weighed, fine earth sample is placed in the 250 ml reaction flask D. Tap A is opened and the moistened bung inserted into the mouth of flask D. Tap B is opened and the manometer reservoir raised until the liquid level in both tubes rises just above the zero graduation. Tap B is closed and the reservoir lowered to bench level. Tap A is then closed and tap C opened to allow in sufficient acid to make the sample fluid. Flask D is carefully shaken so as not to displace the bung and then allowed to stand for several minutes. When the reaction has ceased tap B is opened until the liquid level in the left hand tube is the same as in the graduated tube. Tap B is then closed. More acid is added and the process repeated until a constant level is obtained in the two tubes. The volume of gas is read directly from the graduated tube. Air temperature and barometric pressure must also be recorded.

$$\%\ CaCO_3 \text{ in sample} = \frac{\text{ml } CO_2}{\text{Soil wt.}} \times \frac{\text{Pressure (mm Hg)}}{\text{Temp. } (^{\circ}C + 273)} \times K$$

$$\text{where } K = \frac{273}{760} \times \frac{100}{224} = 0{\cdot}16$$

In cases where other carbonates are exclusively present in a sample, the figure of 100 in the derivation of K may be substituted by the appropriate molecular weight.

7.4 DETERMINATION OF GYPSUM

A number of methods for determination of total sulphates and total sulphur are to be found in the literature (Jackson, 1958; British Standards Institution, 1967; Hesse, 1971). For the majority of agronomic and engineering purposes, however, the estimation of gypsum ($CaSO_4.2H_2O$) provides a satisfactory value for sulphate. The method described here is based on that of Bower and Huss (1948), although the degree to which it is strictly a determination of gypsum is dependent upon the amounts of Na, K and Mg sulphates also present. In this respect the method is analogous to the determination of carbonate content (Section 7.3). Figure 7.3 illustrates the cyclic pathways of sulphur.

20 g air-dry fine earth is weighed into a shaking bottle. 200 ml distilled water are then added and the bottle shaken mechanically for about 20 minutes. The suspension is either filtered or centrifuged and a 20 ml aliquot of the clear solution placed in a 50 ml centrifuge tube. 20 ml acetone are then added and the tube shaken. The tube is allowed to stand for about 10 minutes, after which time a precipitate will have formed if sulphate is present. The tube is then centrifuged for 2 minutes at 2000 r.p.m., the supernatant discarded and the tube inverted and allowed to drain for several minutes onto a filter paper. The precipitate is then dispersed and the tube shaken with 10 ml acetone followed by a repeated centrifuging and draining. Exactly 40 ml distilled water are added from a burette and the precipitate dispersed and allowed to dissolve. 40 ml distilled water will dissolve 0·1 g $CaSO_4.2H_2O$ so that if a soil contains appreciable sulphate the initial extract should be more dilute. Under no circumstances must the soil be oven-dried beforehand.

The electrical conductivity in mmhos/cm is now measured on the dissolved precipitate and a correction applied if the temperature varies from 25°C. A conductivity bridge is used for the determination. The concentration of gypsum must be estimated from the following calibration figures which are plotted to form a graph (Table 7.3).

mmhos/cm at 25°C	$CaSO_4$ m.e./litre
0·121	1
0·226	2
0·500	5
0·900	10
1·584	20
2·205	30·5

TABLE 7.3 *Conductivity and gypsum concentration*

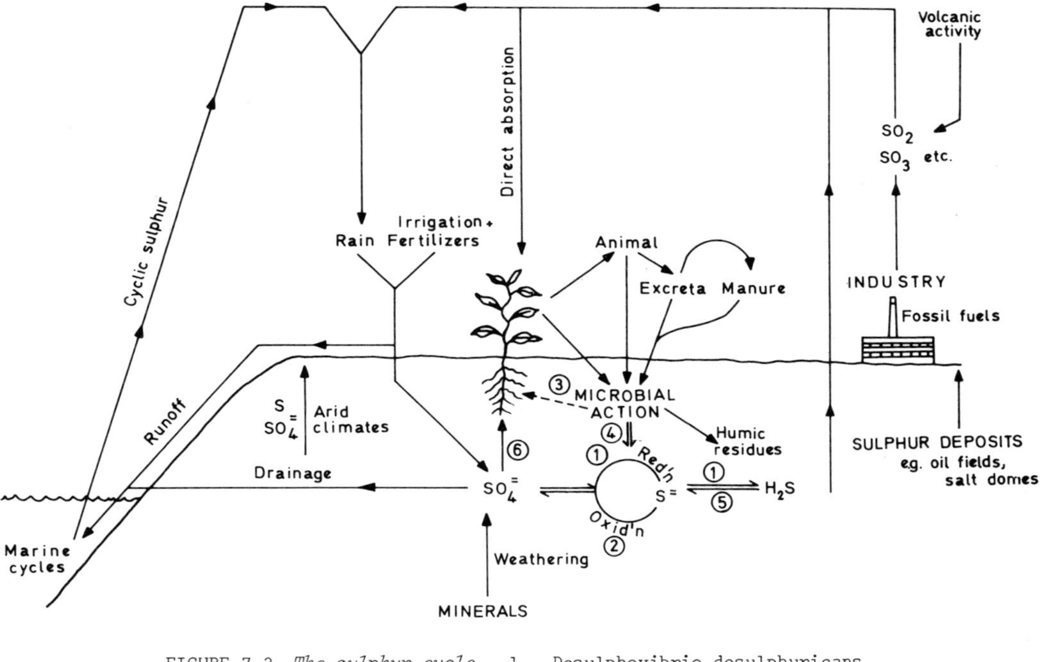

FIGURE 7.3 *The sulphur cycle.* 1. Desulphovibrio desulphuricans *(anaerobic)*; 2. Thiobacillus *(aerobic)*, Chromatium *(anaerobic)*; 3. *transfer of aminoacids*; 4. *mineralisation*; 5. Beggiatoa *(aerobic)*, Chromatium *and* Chlorobium *(anaerobic)*; 6. *uptake by plant.*

$$\text{m.e. gypsum in aliquot } (x) = \text{m.e./litre from graph} \times \frac{40}{1000}$$

$$\text{m.e. gypsum/100 g soil } (y) = \frac{x \times 100}{\text{soil:water ratio} \times \text{ml soil extract used}} = \frac{x \times 100}{10 \times 20}$$

$$\text{\% gypsum} = y \times 0{\cdot}086$$

For determination of the sulphate content of runoff or ground water, the liquid is filtered and a 20 ml aliquot then treated as above. Sulphate, as gypsum, can be read directly from the graph.

A quick field test for sulphates is provided by a weakly acid solution of sodium rhodizonate containing barium ions. In the presence of sulphate the barium is precipitated thus decolourising the red rhodizonate complex. Rhodizonate test papers can be placed onto soil which has first been wetted with dilute HCl.

7.5 DETERMINATION OF SALINITY

Under this heading attention will be paid to the measurement of total water-soluble salts. These are characterised mainly by the anions chloride and sulphate, but include soluble carbonates, hydrogen carbonates (bicarbonates), nitrates, phosphates, borates and silicates. The U.S.D.A. Salinity Handbook (1954) states that a 'saline soil' is one 'having a saturation extract conductivity greater than 4 mmhos/cm at 25°C, a soluble sodium content of less than half the total for the soluble cations and a pH less than 8·5'. The above publication also gives guidelines for the significance of given conductivity readings on saturation extracts (Table 7.4).

Electrical conductivity in mmhos/cm	Plant response to salinity
0-2	Negligible
2-4	Sensitive crops affected (citrus, beans)
4-8	Many crops affected
8-16	Only tolerant crops give satisfactory yield (wheat, grapes, olives)
> 16	Few tolerant crops give satisfactory yield (dates, barley, sugar beet)

TABLE 7.4 *Conductivity readings and salinity responses of plants*

Total salinity is determined rapidly on a conductivity bridge using either a *saturated soil paste*, a *saturation extract*, or a water sample.

A saturated soil paste is made by wetting a 50 g fine earth sample and stirring with a spatula until the soil is just capable of flowing. The paste is allowed to stand for about 1 hour and if still of a satisfactory consistency it is placed in a special conductivity cup, taking care to avoid inclusion of air bubbles. A saturation extract is obtained by spreading the saturated soil paste on a filter paper in a Buchner funnel and employing suction filtration. Alternatively a pressure membrane apparatus is used.

Operating instructions for the conductivity bridge are provided by individual manufacturers and most instruments are battery operated for use in the field. For a saturated soil paste the instrument measures resistance and the operator records the temperature by turning a dial which corrects the instrument to the standard temperature of 16°C. Most instruments record directly in conductance units with different ranges providing varying sensitivity. Reference can be made to U.S.D.A. Soil Survey Manual (1951) and U.S.D.A. Handbook No. 60 (1954) for correction factors and equivalent salt concentration in the soil. The relevant tables have also been reproduced by Hesse (1971).

It is, however, a simple procedure to make one's own set of salt calibration curves with conductivity plotted against varying salt concentrations. Selection of an appropriate range of concentrations is most important for accuracy especially in cases where only small variations in salinity are being measured. For the saturation extract measurement, the temperature has to be corrected to 25°C and, if necessary, the extract diluted until its concentration fits a suitable range on the instrument. This method has the advantage that the extract can be simultaneously tested for soluble nutrient elements.

Providing chloride anion is predominant in the soil, an alternative index of salinity is provided by determination of *water-soluble chloride*. The required reagents include: silver nitrate solution, 0·02*M*; 3·398 g $AgNO_3$ is dissolved in 1 litre H_2O and stored in a dark bottle. Potassium chromate solution: 5 g $K_2Cr_2O_4$ is dissolved in 50 ml H_2O. $AgNO_3$ solution is added until a permanent red precipitate just forms. This solution is then filtered and diluted to 100 ml.

5 g air dry fine earth is shaken mechanically with 25 ml distilled water for 20 minutes. The suspension is then centrifuged and the supernatant filtered through a fine paper into a 25 ml

volumetric flask. The flask is made to volume by washing the soil residue with a small quantity of water. This 25 ml volume can then be titrated with $AgNO_3$ solution using 1 ml $K_2Cr_2O_4$ solution as indicator. A reddish-brown precipitate indicates the end point.

$$\text{m.e. } Cl^-/\text{litre} = \frac{\text{ml } AgNO_3}{\text{ml extract}} \times 20 \quad [x]$$

$$\text{m.e. } Cl^-/100 \text{ g soil} = x \times \frac{100}{\text{soil wt.}}$$

With highly saline soils it is best to titrate a 5 ml aliquot of soil extract with $AgNO_3$ and then multiply the result by 5.

A gravimetric measure of total salinity in an extract or water sample is provided by determination of *total dissolved solids*. A 50 ml portion of liquid sample is evaporated to dryness at 105°C and then ignited at 500°C for 10 minutes to remove traces of organic matter. A silicaware crucible will be a suitable container and all weighings must be as accurate as possible. Results are expressed as p.p.m. solids in solution.

7.6 DETERMINATION OF THE EXCHANGEABLE BASES

Included here are the alkali metals, sodium and potassium and the alkali earths, magnesium and calcium. In soil, the portion of these elements which is held at exchange sites is determined after extraction with neutral *N* ammonium acetate (CH_3COONH_4). This reagent may be prepared from the solid (Appendix C) or by neutralising 57·5 ml glacial acetic acid with 50 ml NH_4OH (0·880 S.G.) and diluting to one litre with distilled water. The pH is adjusted to 7 with either reagent, using bromthymol blue as indicator (Appendix D).

In soils without free salts the extraction procedure is straightforward, while for calcareous, saline or alkali soils experimental procedure, if it is to identify *exchangeable* bases, must differentiate this portion from the total amount extracted. It is, of course, doubtful whether in the latter case a value for exchangeable bases has any validity in nutritional terms since the amount of the elements actually *available* to plants may far exceed this value.

Extraction: The manner of extraction will be determined by available apparatus, the number of samples and the type of soil to be processed. A simple method involves shaking 5-10 g air-dry fine earth with 100-250 ml extractant in a screw-capped bottle for about 15 minutes. The liquid suspension is then filtered or centrifuged to obtain a clear extract. However, the standard method makes use of a leaching column (Figure 7.4). A known weight of carefully subsampled fine earth, usually 5 or 10 g, is packed in the column and then leached with a standard volume of extractant, usually 250 ml. The leachate is then made up to 250 ml with distilled water on account of partial absorption of the extractant by the soil.

Calcareous, saline and alkali soils pose special problems due to the difficulty of separating free salts from the determination of exchangeable ions (U.S.D.A. Handbook, No. 60, 1954). If a soil is saline and the salts are all readily soluble in water they can be leached with water or a water and ethanol solution. When free of salt the soil is leached with ammonium acetate as in a normal extraction. If a soil is calcareous, either the lime must be masked or two extracts collected. In the masking method a 0·2*N* barium chloride-triethanolamine extraction solution is used. This is prepared by diluting 25 ml triethanolamine to

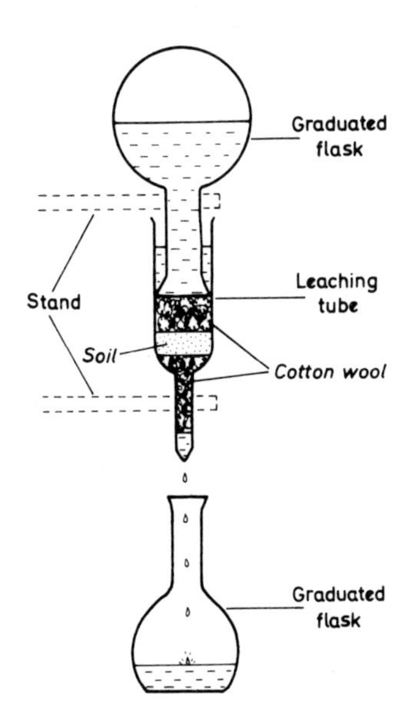

FIGURE 7.4 *A leaching column for cation exchange*

250 ml with distilled water, adjusting the pH to 8·2 with *N* HCl, then diluting to 500 ml with distilled water. This is then added to a solution of 24·4 g $BaCl_2$ in 500 ml distilled water. (Caution: barium chloride is a poison.) The twin-extract method involves leaching the soil with *N* ammonium acetate or *N* NaCl solution. The first 250 ml leachate contains exchangeable Ca (displaced by NH_4 or Na) and a small amount of Ca from $CaCO_3$, whereas the subsequent 250 ml will only contain Ca from solvent action on $CaCO_3$. The two leachates must be clearly labelled A and B respectively. This procedure can easily be adapted to extraction by shaking and filtration if leaching columns are not available.

Standards For calibration purposes *stock standard* solutions of 1000 p.p.m. are prepared as follows:

Na 2·543 g A.R. NaCl dissolved in 1 litre H_2O

K 1·907 g A.R. KCl dissolved in 1 litre H_2O

Mg 10·13 g A.R. $MgSO_4.7H_2O$ dissolved in 1 litre H_2O

Ca 2·497 g A.R. $CaCO_3$ dissolved in 10 ml *N* HCl, boiled to expel CO_2, then diluted with H_2O to 1 litre.

From these solutions *working standards* must be prepared for use with the photometric instruments. These are usually made to a final volume of 100 ml. The maximum concentration of a series of working standards is determined partly by the type of soil being investigated. For non-calcareous and non-saline soils, intermediate standards are made up from 0-20 p.p.m. for Ca, 0-10 p.p.m. for Na and K, and 0-2 p.p.m. for Mg. 0 p.p.m. (zero on the instrument) is represented by a portion of unused extractant while for every 1 p.p.m., 1 ml of stock solution is diluted to 1000 ml with distilled water. If test samples subsequently exceed the value for the strongest standard they can be diluted by a half or a tenth until their values lie within the range of the calibration curve. In general the smaller the range of the calibration curve the more accurate the determination. This is so because the sensitivity is increased and interferences deriving from the solution itself, are diminished the lower the concentration.

Calculation

Values for soil extracts in p.p.m. (*N*) are obtained from calibration curves showing instrument response in relation to concentration standards. Values for exchangeable cations in m.e./100 g soil (*Y*) are obtained as follows:

$$N \text{ p.p.m.} \equiv N \text{ mg/litre} \equiv \frac{N}{\text{E.W.}} \text{m.e./litre}$$

[E.W. Ca 20, Mg 12, Na 23, K 39]

$$\frac{N}{\text{E.W.}} \times \frac{250}{1000} \times \frac{100}{\text{soil wt.}} \times \begin{matrix}\text{any further}\\ \text{dilution of}\\ \text{extract}\end{matrix} \quad Y \text{ m.e./100 g soil}$$

in which 250/1000 represents dilution of leachate to 1 litre. For p.p.m. of an element in soil this would simply be N N soil : extractant ratio (i.e. factor of dilution of soil in extractant) × any further dilution.

Sodium and potassium are determined on a flame photometer using separate filters. The instrument should be set according to manufacturer's instructions then the zero scale reading set and the working standard of maximum concentration used to establish full scale deflection. Intermediate standards may then be used to produce a curve of concentration against percentage transmission (vertical axis). The reading for the test sample is then referred to the calibration graph and the concentration in solution derived. Every few samples it is advisable to check the instrument, ensuring that the scale zero and 100% transmission have not drifted.

If potassium is to be measured as a nutrient element, it is sometimes extracted with dilute (5%) acetic acid. The data are expressed as follows:

$$\text{mg available K/100 g soil} = \frac{\text{p.p.m. from curve}}{10} \times \text{soil:extractant ratio}$$

To convert K to K_2O the above figure is multiplied by 1·20. Table 7.5 shows the figures from Gilchrist-Shirlaw (1967).

When atoms, molecules or ions are subjected to energy-excitation as in a flame, energy is first absorbed by shifts of electrons to positions more distant from the nucleus of the atom. These electrons then return at least partially to the ground state, at which instant the absorbed energy is emitted as electromagnetic radiations or colour. The wavelength of this light corresponds to the amount of energy involved in the respective electron shifts. The *flame photometer* monitors the light emitted when a sample extract is sprayed into a town or

Potash status	mg K_2O/100 g soil
Very low	0-6
Low	6-10
Medium	10-18
Medium-high	18-28
High	28-35
Very high	> 35

TABLE 7.5 *Guide to potash status*

natural gas/air flame. The light, characteristic of a given element, is isolated by the appropriate gelatine filter and measured by a photoelectric cell coupled to a galvanometer.

Calcium may also be determined on a flame photometer providing phosphate has been eliminated. A 25 ml aliquot of extract is placed in a centrifuge tube with 2·5 ml saturated ammonium oxalate. Calcium oxalate precipitates. The tube is heated in a water bath at 90°C for 15 minutes and is then cooled and centrifuged for 1 minute at 2000 r.p.m. The supernatant is discarded. The compacted precipitate is dissolved by addition of 1 ml 2*N* HC*l* and warming to 60°C. *N* potassium permanganate (3 g $KMnO_4$ in 100 ml H_2O) is added dropwise until a pink colour has established. A few drops of hydrogen peroxide are then added until the solution is colourless. It is then made up to 100 ml with water, a dilution of four times the original aliquot. This solution is then tested as for Na and K but with the Ca-filter inserted.

Calcium and magnesium are usually determined on an atomic absorption spectrophotometer (McPhee and Ball, 1968). The procedure for the two elements is the same apart from the concentration of standards used and selection of the appropriate wavelength, the atomic line for Ca being 422·7 nm and for Mg, 285·2 nm.

The fact that anions may form stable compounds with a cation in an air-acetylene flame can be overcome by adding an excess of a 'competing' cation otherwise known as an *internal standard.* A 5% w:v solution of lanthanum is used for this purpose (e.g. 9 g S.P. $LaCl_3$ dissolved in 100 ml water). All standards should contain 1% La (20 ml La solution in final 100 ml). Standards are then measured on the instrument in order to prepare calibration curves. 2 ml of the lanthanum solution is then added to 8 ml soil extract in an acid-clean, glass beaker. After mixing, this test solution is measured and p.p.m. Ca or Mg in solution obtained from the curves. In the *atomic absorption spectrophotometer* the aerosol sample is burned in the higher temperature air and acetylene flame, in which a proportion of free atoms are formed. Pulsed light from a hollow cathode lamp emitting the wavelength of the element under investigation, passes through the flame and a monochromator, and the intensity of the signal (e.g. Ca at 422.67 nm) is reduced in proportion to the amount of substance in the flame.

Magnesium may, in addition, be determined as a yellow complex using a spectrophotometer or colorimeter. The required reagents are as follows: sodium hydrogen sulphite solution 0·5% w:v in water, stored in a dark bottle; polyvinyl alcohol 1% w:v in water. The solid dissolves on shaking and froth disappears on gentle heating. Compensating solution, is made up from 3.3 g $CaCO_3$, 0·36 g $MnCl_2$, 0·74 g $Al_2(SO_4)_3$ and 0·60 g Na_3PO_4 which are dissolved in 500 ml water to which 100 ml conc. HCl has been added. The solution is then diluted to 1 litre. Thiazole Yellow, 0·04% w:v in water stored in a dark bottle. This reagent should be specifically for Mg determination (e.g. Titan yellow, Michrome No. 163 supplied by E. Gurr Ltd., London). Sodium hydroxide, 400 g/litre (10*N*).

A 10 ml aliquot of soil extract is pipetted into a 25 ml flask. 5 ml of a mixture of equal parts of sodium hydrogen sulphite, polyvinyl alcohol and interference-compensating solution, are added, and the flask swirled. 1 ml Thiazole Yellow and 4 ml NaOH are added and the contents diluted to 25 ml with *N* ammonium acetate. The yellow colour is stable for about 1 hour and the range of measurement is from 0-3 p.p.m. Optical density is measured at 540 nm (green or yellow-green filter). Colours are similarly developed using standards of 0.5, 1, 2 and 3 p.p.m. Mg. A calibration curve is drawn with concentration plotted against optical density (vertical axis) and the reading for the diluted soil extract referred to this (Gilchrist-Shirlaw, 1967; Hesse, 1971).

Calcium and magnesium can also be determined together in the same extract by complexometric titration with EDTA. The required reagents include, 0·02*N* EDTA solution: 3.3 g ethylene diamine tetra-acetic acid disodium salt dissolved in 1 litre H_2O. This solution should be standardised against 0·02*N* $CaCl_2$ (1·0 g $CaCO_3$ dissolved in just sufficient *N* HCl and made to 1 litre). 0·02*N*(or *N*/50) EDTA is available in ampoules ready for dilution to 500 ml. Murexide indicator: 0·2 g ammonium purpurate is mixed with 10 g NH_4Cl and 40 g NaCl, and finely ground in a mortar, *or* Eriochrome Blue Black R (Calcon): 0·2 g in 50 ml methanol - stable for up to 2 weeks. Alkali buffer (pH 10): 33·75 g NH_4Cl dissolved in 100 ml water added to 214 ml 0·880 S.G. NH_4OH and diluted to 500 ml with water. Eriochrome Black T indicator: 0·2 g in 50 ml methanol containing 1 g hydroxylamine hydrochloride - stable for about 1 week.

For calcium, 20 ml of the single leachate (or leachates A and B referred to earlier) is placed in a conical flask(s). 2 ml 10% NaOH are added and about 1 g murexide indicator. With EDTA in the burette, titration is carried out until the colour changes from pink to pale violet. The end point may be compared with that of a blank (NaCl). This is a difficult colour change to estimate, the alternative indicator giving a colour change from pink to blue. It is helpful if a specimen of leachate and reagents alone is used as a standard to indicate any departure from the initial colour occurring in the course of titration.

$$1 \text{ ml } 0{\cdot}02N \text{ EDTA} \equiv 0{\cdot}02 \text{ m.e. Ca} \quad (X)$$

For magnesium, 20 ml of the leachate(s) is placed in a conical flask. 5 ml alkali buffer and 5 drops Eriochrome Black T indicator are added. This is again titrated with the EDTA until the colour changes from purple to blue. The colour change may again be compared with that of a blank.

$$1 \text{ ml } 0{\cdot}02N \text{ EDTA} \equiv 0{\cdot}02 \text{ m.e. Ca + Mg} \quad (Y)$$

For a single leachate Ca is given directly and Mg is derived by difference. For the twin leachate method:

$$X\text{A} - X\text{B} = \text{m.e. exchangeable Ca} \quad (Z)$$

$$Y\text{A} - Y\text{B} - Z = \text{m.e. exchangeable Mg}$$

7.7 DETERMINATION OF EXCHANGEABLE HYDROGEN AND CATION EXCHANGE CAPACITY

Cation exchange capacity in m.e./100 g at pH 7 can be derived by summation of the individual values for the principal bases (Section 7.6) together with that of exchangeable hydrogen. Alternatively it can be determined directly by first leaching the soil with neutral *N* ammonium acetate and then displacing the ammonium with NaCl solution and measuring the amount of ammonium ion previously adsorbed.

Exchangeable hydrogen The ammonium acetate leachate from the exchangeable bases determination is used. From the original 250 ml, a 25 ml aliquot is pipetted into a conical flask and

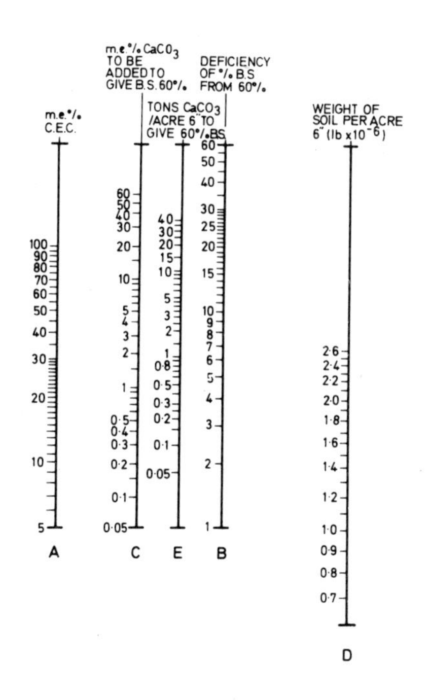

FIGURE 7.5 *Nomogram for determination of lime requirement (after Metson, 1961)*

titrated back to pH 7 using 0.2*N* NH_4OH, and bromthymol blue as indicator.

$$\text{Exchangeable H} = (\text{ml } NH_4OH \times 0.2 \times \frac{250}{25} \times \frac{100}{\text{soil wt.}}) \text{ m.e./100 g}$$

For the separate determination of C.E.C. the soil in the leaching column is leached with about 100 ml 60% methanol which removes excess ammonium acetate from around the soil particles without displacing adsorbed NH_4^+. (This assumes that exchangeable bases have already been determined and will not be necessary if fresh soil is used.) The soil is then leached with 250 ml *N* NaC*l*. The leachate is made up to 250 ml with distilled water and a 20 ml aliquot steam distilled by the same procedure as used in the Kjeldahl determination of nitrogen (Section 7.13). After distillation, the contents of the receiving flask are titrated with 0·05*N* HC*l* so that:

$$\text{C.E.C.} = (\text{ml HC}l \times 0.05 \times \frac{250}{20} \times \frac{100}{\text{soil wt.}})\ \text{m.e./100 g}$$

When C.E.C. is known and also the figure for total exchangeable bases (T.E.B.), the percentage base saturation (B.S.) can then be derived. Given the latter, it is possible to estimate the *lime requirement* of a soil and Figure 7.5 illustrates a nomogram used for this purpose (Metson, 1961). An example will illustrate its use. A soil has C.E.C. of 35 m.e./100 g, a B.S. of 20%, i.e., 40% below the critical 60% level. The soil weighs approximately $1{\cdot}6 \times 10^6$ lb/acre to a depth of 6 inches. 35 on scale A is aligned with a ruler to 40 on scale B which gives a value of 14 on scale C. This value of 14 is then aligned with 1·6 on scale D to give 5 tons per acre on scale E. (1 ton = 1016 kg; 1 lb $\times 10^6 = 0{\cdot}45 \times 10^6$ kg; 1 acre 6 in = 4×10^3 m^2 15 cm). It is clear that for accurate use of the nomogram an estimate must be made of the acreage weight of a soil which will depend on its texture, the higher values referring to 'heavier' soils. Another method is given in Bear (1964).

7.8 DETERMINATION OF COPPER, LEAD AND ZINC

Measurement of trace elements presents the analyst with his greatest challenge, as many such substances are present in laboratory tap water, on hardware and glassware and even in reagents, so that collectively these sources may exceed the concentration in the soil sample. Colorimetric methods for determination of trace elements are found in Jackson (1958), Black *et al.* (1965) and Hesse (1971), but for very low concentrations atomic absorption spectroscopy is a superior and quicker method.

Copper and zinc are both micronutrients which function in enzymes. Lead is non-essential but is accumulated in plants especially in acid soils. All three elements are, however, linked by their presence in industrial fallout and other forms of contamination. Zinc in soils is the most soluble (as zincate) while Pb and Cu are usually held by organic matter and are therefore more subject to accumulation. Apart from the consequences of Cu and Zn deficiency which affects animals dependent upon herbage, interest centres on locally high levels of these elements in soils and inland waters, and its possible consequences to human health (Warren, 1964). Combustion residues from hydrocarbons contain ten times the average concentration of Pb in the earth's crust. Recent studies of garden soils (Purves, 1966) show that urban areas may have at least five times the normal level of Cu and Pb. Riverside pasture soils in a Welsh valley had 1780 p.p.m. Pb, 320 p.p.m. Zn and 22 p.p.m. Cu more than fifty years after cessation of mining activities upstream, while values of 58 p.p.m. Pb, 60 p.p.m. Zn and 8 p.p.m. Cu were obtained for a local uncontaminated soil (Davies, 1968).

For estimations of *total* Cu, Pb and Zn, the soil must be digested with HF and $HClO_4$ with evaporation to dryness until the residue dissolves on addition of HCl. 1 g 0·15 mm soil (sieved through nylon meshes) is accurately weighed into a crucible. As a substitute for the highly-priced platinum crucible, PTFE beakers may be used providing the temperature does not exceed 300°C. Polythene covers with radial ribs to allow evaporation, should be used. 20 ml HF and 2 ml $HClO_4$ are added and the mixture heated carefully on a sand tray or steam bath in a fume cupboard. Organic matter is thus oxidised and silica vapourised as silicon tetrafluoride.

When dry, 20 ml 6*N* HCl is added, the crucible boiled briefly and when all residue has dissolved, the contents are cooled, transferred to a 100 ml volumetric flask and diluted to volume with redistilled water. One alternative to this method is Na_2CO_3-fusion (Section 7.9 and see Bear, 1964).

For *available* Cu, Pb and Zn, a standard method is to shake a 5 g air-dry, fine earth sample (sieved through nylon) for about 30 minutes with a 30 ml mixture of HCl and H_2SO_4 (50 ml *N* HCl and 25 ml *N* H_2SO_4 diluted to 1 litre with water). Alternatively 0·1*N* HCl or 0·5*N* HNO_3 has been used, while Morgan's Reagent (Section 7.10) is most suitable for calcareous soils. The suspension is filtered through a Whatman No. 42 paper into a 50 ml volumetric flask and made to volume with further extractant (Isaac and Kerber, in Walsh, 1971).

Standards and samples should contain approximately 1% La in their final volume (Section 7.6). Stock standard 1000 p.p.m. solutions are made up as follows from analytical grade reagents:

Cu   1·964 g $CuSO_4.5H_2O$ dissolved in 1 litre H_2O

Pb   1·998 g $Pb(NO_3)_2$ dissolved in H_2O with 20 ml conc. HNO_3, then diluted to 1 litre

Zn   4·549 g $Zn(NO_3)_2.6H_2O$ dissolved in 1 litre H_2O

Working standards are prepared to cover suitable concentration ranges, usually 0-1 or 0-10 p.p.m. Copper is measured at 324·7 nm, lead at 283·3 nm and zinc at 213·9 nm. Detection limits by this method are approximately 0·005 p.p.m. for Cu, 0·03 p.p.m. for Pb and 0·002 p.p.m. for Zn. The concentration of the test solution is obtained from the standard curve.

$$\text{p.p.m. element in soil} = \text{p.p.m. in solution (from curve)} \times \frac{\text{ml original dilution of extract}}{\text{soil wt.}} \times \text{any further dilution of extract}$$

For assay of these elements in water samples the prepared standards should be diluted from stock with dilute HNO_3 and both standards and samples should contain 1% La. The water samples should be collected in acid-clean plastic bottles and will normally have to be concentrated. A 500 ml aliquot is transferred to any convenient dish and 15 ml conc. HNO_3 added. This is then evaporated to about 15 ml and transferred to a 50 ml acid-clean volumetric flask and made to volume with redistilled water.

Although this section has been concerned with the determination of Cu, Pb and Zn, other elements are also present in the extracts and in addition to the exchangeable bases these include Al, Fe, Mn, Ni, Co, Cr and B. Determination of these and other rarer elements can also be carried out by absorption spectroscopy. The procedure above remains essentially the same although changes in gas mixtures and burner-type are necessary.

7.9 DETERMINATION OF IRON

Iron in soil may be present in primary minerals, as hydrated oxides coating and cementing particles or mineral aggregates, and as insoluble salts, oxides and organic complexes in the clay fraction. Iron may also be present in the ferrous or ferric form which further complicates the business of extraction, rendering iron in solution no less easy to define than is phosphorus (Section 7.10). Iron is also an essential micronutrient element. A deficiency of iron leads to chlorosis in plants and to anaemia in animals and Man. Deficiency may be due to low reserves but more commonly is the result of interference by high concentrations of other elements such as Ca, P, Zn and Cu. Iron is responsible for the majority of colours exhibited by soil profiles and its mobility in soil is intimately related to soil aeration and soil acidity. The latter influences the types of organic compounds present in the soil. Depending upon which fraction of iron is required, an appropriate extractant has to be chosen. A dilute solution of a weak acid such as citric is most suitable for the 'available' form of the element.

Extraction of total iron Iron may be measured after perchloric acid digestion as for total P in Section 7.10, after HF/$HClO_4$ treatment as in Section 7.8, or after fusion with sodium carbonate as described here. About 1·0 g, 0·15 mm oven-dry soil is weighed accurately into a platinum or silicaware crucible and mixed with about six times the quantity of anhydrous sodium carbonate. The mixture is then covered with a layer of carbonate powder, placed in a furnace and heated to 1200°C. When liquid, the melt is swirled, then heated for a further 15 minutes. The melt is finally swirled so that a thin layer forms on crystallising. After cooling, the crucible is placed in a beaker and covered with water. The beaker is warmed in a water bath until the crucible contents have dispersed. The crucible and lid are rinsed into the beaker. The contents of the beaker are then acidified with about 10 ml conc. HCl and 10 ml 60% $HClO_4$. When reaction has died down a glass cover is placed over the beaker and the contents evaporated to the fuming stage. Heating is then continued for a few minutes to dehydrate the silica. The solution is then cooled, about 25 ml warm water added, and the solution filtered through a Whatman No. 41 paper into a 100 ml volumetric flask. The silica residue is washed with 0·5*N* HCl and the flask made to volume with 0·5*N* HCl. To produce a melt it may be simpler to use a test tube in a bunsen flame. Total Fe is measured by either method A or B.

Extraction of free iron oxides Iron oxides are brought into solution by sodium dithionite in a citrate solution. Reagents include: citrate buffer solution 10.5 g citric acid monohydrate and 147 g sodium citrate (tribasic) dissolved in 1 litre H_2O. Sodium dithionite, powder. Perchloric acid, 60% (Coffin, 1963).

A 1 g sample of air-dry 0·15 mm soil is weighed into a 50 ml centrifuge tube. 20 ml citrate buffer and 1-2 g sodium dithionite are then added. The tube is stoppered and shaken periodically in a water bath at 50°C for 30 minutes, or the soil and extractant can be shaken overnight without heating. The tube is then centrifuged (or filtered through Whatman No. 1 paper) and a portion of supernatant (e.g. 1 ml) digested in a tallform beaker with 2 ml $HClO_4$ until almost dry. The residue is dissolved in 0·5*N* HCl and made to volume in a 100 ml flask.

Alternatively, the following is a quicker but more drastic extraction. 1 g soil is heated with 10 ml conc. HCl and 0·5 ml conc. HNO_3. After a few minutes an equal volume of water is added and the suspension filtered. The residue is washed twice and the filtrate diluted to 100 ml.

Determination of iron - Method A 1 ml of the diluted iron extract is placed in a centrifuge tube. 1 drop of thioglycollic acid, 1 ml NH_4OH (diluted 1:3·3) and 0·5 ml citric acid (20% w:v) are added successively, followed by water to make a final volume of 10 ml. A violet colour develops and an aliquot of the solution is measured on a spectrophotometer at about 570 mm (or colorimeter with yellow filter). A calibration curve is constructed from 0-10 p.p.m. Fe using standard Fe solution in flasks followed by the above reagents. The standard 100 p.p.m. Fe is made by dissolving 0·7022 g ferrous ammonium sulphate in 1 litre H_2O containing 5 ml conc. H_2SO_4. 1 ml of this solution is diluted to 100 ml for each 1 p.p.m. (Cornwall, 1958; Tintometer Ltd.).

$$\text{p.p.m. Fe } (x) = \text{p.p.m. from curve} \times \frac{\text{ml final volume}}{\text{ml aliquot from extract}} \times \text{soil : extractant ratio (100)}$$

$$\%Fe_2O_3 = \frac{x \times 1\cdot43}{10^4} \qquad \left[\text{where } 1\cdot43 = \frac{\text{Atomic weight } Fe_2O_3}{\text{Atomic weight } Fe_2}\right]$$

In colorimetric methods such as this, the soil extracts and standards being measured, are coupled with another compound (frequently organic) to yield a coloured complex. The intensity of the colour should depend upon the concentration of the substance and thus obey Beer's law. The instrument used for this determination is either a lamp or grating *spectrophotometer*, or a *colorimeter* with filters, both of which measure the absorbance which occurs as light passes through a standard-sized glass tube or cell containing the coloured solution.

Determination of iron - Method B A 1 ml aliquot of soil extract is pipetted into a 100 ml volumetric flask. About 70 ml distilled water are added, followed by these reagents: 5 ml conc. HCl, 1 ml 30 vol H_2O_2, 5 ml 15% w:v potassium thiocyanate (30 g A.R. KCNS dissolved in 200 ml H_2O - this reagent is deliquescent and may need drying in an oven at 105°C). These reagents must be treated with the greatest care especially as KCNS is highly poisonous. They must be transferred to the flask using a safety-pipette which is suitably graduated. The flask is made to final volume with water. The red colour which forms, must be allowed about 30 minutes to develop fully. Optical density is then measured at 480 nm (blue or blue-green filter). Standardisation and calculation are as for method A.

Field test for ferrous and ferric iron Freshly collected soil is placed in separate beakers or in the cavities of a white tile and moistened with dilute (*N*) HCl. Potassium thiocyanate solution (10% w:v) is added to one of the samples and *ferric* iron is indicated by a red colouration. Potassium ferricyanide (hexacyano-ferrate III) solution (0·5% w:v) is added to the other sample and *ferrous* iron is indicated by a blue colouration. These colours may be made more evident on the surface of the soil by adding a white powder e.g. kaolin.

7.10 DETERMINATION OF PHOSPHORUS

Phosphorus rarely comprises more than 0·1% of the soil and only a small amount of plant-available inorganic orthophosphate ion is present at any time. The bulk of phosphorus is usually unavailable, being fixed either as Ca phosphate at high pH or as *Al* and Fe phosphates at low pH. Much phosphorus is stored in organic matter as inositol hexaphosphate (Figure 7.6). Measurement of total phosphorus is therefore of limited value in determining phosphorus availability except that low concentration usually indicates deficiency.

It is difficult to recommend one particular extractant for *available* phosphorus because values obtained using different reagents on the same soil will vary. The effectiveness of an extractant for available nutrients is also often limited to a particular range of soil conditions. As with all extractions it is largely a question of what the values mean once they have been determined. Many extracting solutions have been used for obtaining the 'more soluble forms' of the element and as such, some have undoubtedly overestimated what is available to the plant. The following experimental details

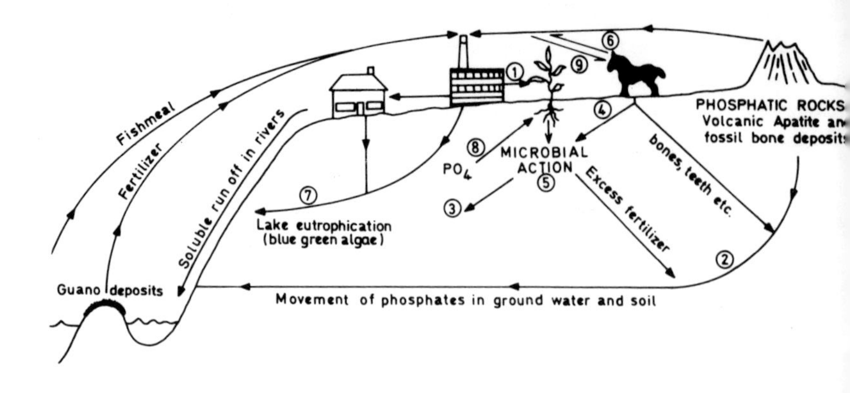

FIGURE 7.6 *The phosphorus cycle. 1. phosphate fertilisers, NPK, bone meal etc.; 2. leaching of dissolved phosphates; 3. immobilisation (fixation); 4. urea and dung; 5. bacterial domain with phosphatising species; 6. animal remains e.g. bone, hoof and horn; 7. domestic and industrial wastes e.g. synthetic detergents; 8. available phosphates (especially via mycorrhizae); 9. concentrated animal feed.*

(Table 7.6) are provided by Boyd (1962) who investigated the relationship between available P and crop response for a range of soil types.

Olsen's reagent	shaken for	30 min	1:20
Morgan's reagent	" "	15 min	1:5
1% citric acid	" "	24 hr	1:10
0·5% acetic acid	" "	1 hr	1:40
0·3 N HC*l*	" "	1 min	1:2·5
0·01 N $CaCl_2$	" "	15 min	1:2

TABLE 7.6 *Extractants for 'available' phosphorus*

The list is in approximate order of effectiveness for the range of soils studied by Boyd and the final column refers to the soil:extractant ratio. Olsen's reagent is 0·5*N* $NaHCO_3$ at pH 8·5 while Morgan's reagent (of which there are modifications) consists of 0·734*M* sodium acetate and 0·518*M* acetic acid at pH 4·5 prepared by dissolving 100 g sodium acetate in water, adding 30 ml glacial acetic acid and diluting to 1 litre.

Olsen's bicarbonate has been adopted for general use by the Agricultural Advisory Service in the U.K. (M.A.F.F., 1973). Further discussion is offered by Larsen (1967).

Extraction with hot water has also been found satisfactory especially for circumneutral soils. A soil:extractant ratio of 1:10 or 1:20 can be considered satisfactory, with shaking for at least 15 minutes. Either method A or B can be chosen for the determination.

For the extraction of *total phosphorus* 1-2 g air-dry 0·2 mm soil is weighed accurately into a 250 ml tallform beaker and 20 ml conc. HNO_3 added. The beaker is covered and gently heated to oxidise organic matter. 10 ml 60% $HClO_4$ is added and the mixture digested with a cover until white fumes appear. Heating is continued for a further 5-10 minutes to dehydrate silica. The solution is cooled, about 25 ml warm water added and the solution filtered through a Whatman No. 41 paper into a 100 ml volumetric flask. The silica residue is washed with 0·5*N* HCl. Either method A or B can be chosen for the determination.

For the extraction of *total organic and total inorganic phosphorus* 2 g 0·2 mm air-dry soil is weighed into a crucible and ignited in a furnace for 1 hour at 400°C. The ignited soil (X) is then transferred to a 100 ml centrifuge tube and a further 2 g soil sample (Y) placed into another tube. 10 ml conc. HCl is added to both tubes, which are then placed in a boiling water bath for about 10 minutes followed by the addition of a further 10 ml conc. HCl. The tubes are then cooled and about 1 hour later, 50 ml water added and the tubes centrifuged. The supernatant is decanted into labelled 250 ml volumetric flasks and made to volume with distilled water. Phosphorus is best determined as soon as possible by method A below so that hydrolysis is minimised (Hesse, 1971).

Organic-bound P = p.p.m. X - p.p.m. Y

Note: It is best not to use plastic or polythene vessels for work on phosphorus and all glassware should be rinsed in dilute HNO_3 after cleaning in detergent.

Method A Required reagents include, standard phosphate solution: 0·2195 g A.R. KH_2PO_4 dissolved in 500 ml H_2O and 25 ml 3·5*M* H_2SO_4. When diluted to 1 litre this makes 50 p.p.m. phosphorus. Ammonium molybdate solution: 50 g A.R. $(NH_4)_6 Mo_7O_{24}.4H_2O$ dissolved in 400 ml 5*M* H_2SO_4 and 500 ml H_2O and diluted to 1 litre. Sulphuric acid 0·5*M*: 28 ml conc. acid per litre. Stannous chloride solution: 10 g $SnCl_2$ dissolved in 25 ml conc. HCl. 1 ml of this solution is diluted to 200 ml. Butan-2-ol reagent.

A 10 ml aliquot of the soil extract is pipetted into a separating funnel marked at 20 ml. 5 ml ammonium molybdate solution is added and the liquid diluted to the 20 ml mark with further extractant. 10 ml butan-2-ol is added and the funnel shaken for 2 minutes. The aqueous layer is discarded by opening the tap, and the organic layer washed with 10 ml 0·5*M* H_2SO_4. To this, 15 ml diluted $SnCl_2$ solution is added and the funnel shaken for 1 minute. The aqueous layer is again carefully discarded and the organic phase transferred to a 50 ml volumetric flask. The funnel is rinsed into the flask with ethanol and the flask then made to volume with ethanol. 40 minutes later the solution is measured at 730 nm on a spectrophotometer or colorimeter with red filter. A standard curve must be constructed to cover the range 0-1 p.p.m. phosphorus. The soil extract is diluted to fit the curve if it is too concentrated. Calculation is as given for method B.

Method B Required reagents include standard phosphate solution as for method A. Ammonium vanadate solution: 2·345 g anhydrous ammonium vanadate are dissolved in 400 ml hot water. 17 ml 60% perchloric acid is added and the solution then diluted to 1 litre. Ammonium molybdate solution: 25 g ammonium molybdate dissolved in 400 ml water at 50°C. This is cooled, filtered if necessary, diluted to 500 ml and stored in a dark bottle.

A 10 ml aliquot of soil extract is pipetted into a 50 ml volumetric flask. 8 ml, 60% perchloric acid is added and the solution diluted to about 30 ml with water. 5 ml ammonium vanadate solution is then added and any silica precipitate is filtered off. 5 ml ammonium molybdate solution is then added and the flask diluted to volume with distilled water.

Appropriate volumes of 50 p.p.m. P solution are placed into four, 50 ml volumetric flasks to give 1, 5, 10 and 20 p.p.m. P when diluted to 50 ml. The above reagents are added to these flasks, and a further flask (0 p.p.m. P) given reagents and distilled water only. A yellow colour takes about 30 minutes to develop fully and is then measured at 470 nm on a spectrophotometer or with a colorimeter using a blue filter. From the standards, a curve is drawn of concentration against absorbance. The soil extract is then compared with this curve and is diluted if too concentrated. The yellow colour is only stable for a few hours. Greater sensitivity may be achieved by reducing the wavelength (Jackson, 1958).

If phosphorus concentrations are to be related to a weight of soil they must be multiplied by certain factors as follows:

$$\text{mg P/100 g soil} = \frac{\text{measured p.p.m.}}{10} \times \frac{\text{ml final volume}}{\text{ml aliquot from extract}} \times \text{soil:extractant ratio}$$

To convert P to P_2O_5 the above figure is multiplied by 2·29.

The figures from Gilchrist-Shirlaw (1967) are shown in Table 7.7.

Phosphate status	mg P_2O_5/100 g soil
Very low	0-1·4
Low	1·5-3·0
Medium	3·1-6·0
Medium-high	6·1-10·0
High	10·1-20·0
Very high	> 20·1

TABLE 7.7 *Guide to phosphate status*

An almost universal feature of adding phosphorus to soils is the relatively poor response of crops in relation to the amount applied. This is almost entirely due to immobilisation within the soil. The *phosphorus fixation capacity* of a soil can be estimated by the following procedure: 1 g air-dry fine earth sample is shaken with 20 ml, 20 p.p.m. standard P solution for up to 2 hours. The suspension is then centrifuged at 2000 r.p.m. for 1 minute and phosphorus determined by either method A or B. The change in concentration of P in the standard solution indicates the fixation capacity, and this can be expressed as mg P/100 g soil.

Phosphorus may also be measured in runoff or ground water samples by either methods A or B, the results being expressed as p.p.m. in the original sample. The reader may be aware that electrical probes are available to test for a wide range of substances in ionic form, including phosphate. While a rapid method, this does require careful calibration in order for the accuracy to match that of the colorimetric method. Probes have been mainly used for monitoring ionic substances in water samples.

7.11 DETERMINATION OF ORGANIC MATTER BY IGNITION

The ignition or dry combustion method ideally involves the use of an electrically heated furnace, but for demonstration purposes a bunsen burner will suffice. Organic substances are volatile when heated in an ordinary atmosphere so that the method can satisfactorily be used to indicate total organic matter and total organic carbon (Ball, 1964). The loss-in-weight principle is a simple one to follow, yet the method suffers from being non-specific and provides an index of organic matter content rather than a meaningful absolute value. For instance, a small amount of ash always remains after burning pure organic matter and there are other sources of weight-loss besides that deriving from organic matter. Sulphides may be converted to sulphur dioxide. Some clay minerals lose structural water (450-600°C) and calcareous soils lose carbon dioxide from carbonates (850-1000°C). Furthermore, no distinction is possible between elemental carbon and organic matter in various stages of decay. As a general rule, soils with appreciable quantities of charcoal or soot, which have a high proportion of hydrated clays and which are calcareous, should be treated cautiously. If loss in weight is plotted against temperature of ignition some of these errors may be checked and this procedure is recommended in the questionable cases. Comparisons between samples on similar parent materials will tend to be more reliable, but providing ignition temperature does not exceed about 400°C, a number of errors can be reduced. Unavoidably, however, hydroxyl ions and water of hydration will be lost even at these low temperatures and so for this and other reasons, percentage weight loss on ignition is usually more than the true percentage of organic matter. Although air-dry samples may often be ignited without sizeable errors, samples are best oven-dried overnight at 105°C to remove hygroscopic moisture. The loss in weight after ignition can then be standardised as the weight lost after the 105°C moisture loss. Figures which can be expected range from 5-12% for arable land and around 15% for some horticultural soils and soils under permanent pasture, while values of over 50% are commonly obtained from peats and mor humus.

Up to 10 g oven-dry fine earth (< 2 mm) is placed into a weighed and labelled porcelain or silicaware crucible. The crucible and contents are then weighed accurately and placed in an electric muffle furnace. Crucibles should be handled with a pair of furnace tongs and it is advisable to practice using these on an empty crucible beforehand! Where a batch of samples is being treated it is advisable to make a diagram of the placement of crucibles in the furnace or ensure that the marker for labelling is heat-resistant. Ignition is carried out in the range 375-425°C for a maximum of 24 hours. An 8-hour period is usually adequate and the shorter time helps to reduce inorganic weight losses. After ignition the samples are allowed to cool in a desiccator so that they do not gain hygroscopic moisture before reweighing. If desired, the residue from low temperature ignition may be subjected to chemical analyses.

$$\text{\% wt. loss or organic matter index} = \frac{\text{initial wt. sample} - \text{final wt. sample}}{\text{initial wt. sample}} \times 100$$

Since organic matter commonly contains around 58% carbon, organic carbon content will approximate to the percentage weight loss on ignition multiplied by the factor 58/100.

7.12 DETERMINATION OF ORGANIC CARBON

The method described is based on that of Walkley and Black (1934). Soil is digested with a mixture of chromic and sulphuric acids, utilising the latter's heat of dilution. Non-reliance on externally applied heat means that 90-95% of elementary carbon remains unattacked. Organic carbon is converted to carbon dioxide by the chromic acid which itself is thereby reduced to chromic sulphate. Since the amount of chromic acid consumed is proportional to the amount of organic matter in the soil, it is necessary to determine the former by titrating the unreacted excess of chromic acid with a standard ferrous solution.

High soil concentrations of chloride, ferrous iron, sulphides and the higher oxides of manganese cause interference but can be eliminated by pretreatment (Jackson, 1958). The presence of chlorides leads to high results, so the affected sample should be leached with water until washings give no turbidity on addition of $AgNO_3$ solution. Sulphides also lead to high results and the affected samples should be treated with approximately 2*N* H_2SO_4 followed by a leaching with water (B.S.I., 1967). Gleyed soils should be well broken up and oven-dried overnight at 105°C to convert ferrous iron to the ferric form.

The reagents required for this procedure are normal potassium dichromate: 49.035 g $K_2Cr_2O_7$ are dissolved in water and diluted to 1 litre; ferrous solution: 0·5*N* ferrous ammonium sulphate prepared by dissolving 196·1 g $Fe(NH_4)_2(SO_4)_2 \cdot 6H_2O$ in 800 ml water with 20 ml conc. H_2SO_4 and diluting to 1 litre, or alternatively normal ferrous sulphate in which 278·0 g $FeSO_4 \cdot 7H_2O$ and 15 ml conc. H_2SO_4 are dissolved in 1 litre of water (ferrous sulphate solution is the less stable and should be kept tightly stoppered); diphenylamine indicator: 0·5 g diphenylamine is dissolved in 20 ml water and 100 ml conc. H_2SO_4; alternatively 0·25 g sodium diphenylamine sulphonate is dissolved in 100 ml water.

Air-dry soil is ground to pass a 100-mesh sieve. About 1 g is transferred to a 500 ml conical flask. 10 ml (20 ml if an organic surface sample) *N* $K_2Cr_2O_7$ is added, followed by 20 ml 95% H_2SO_4. The flask is swirled gently for about a minute ensuring that no soil escapes contact with the reagents. The flask is then allowed to stand for 20-30 minutes on a heat-proof surface (e.g. an asbestos mat). The solution is diluted with 200 ml distilled water and 10 ml 85% orthophosphoric acid (H_3PO_4). To this mixture is added 0·2 g solid NaF and 1 ml diphenylamine indicator. This is then titrated against the standard ferrous solution (in burette). Initially, a dull green chromous colour is observed; this will turn blue, while the end point (quite sharp) is a bluish green. If the end point is over-shot a further 0·5 ml *N* $K_2Cr_2O_7$ is added and the titration continued. A blank titration is run using the same procedure but without adding soil. This standardises the ferrous solution against the chromic acid.

The oxidisable matter by this procedure only constitutes about 77% of that removed by ignition, so it is necessary to multiply the milliequivalent weight of carbon by this recovery factor in order to derive a weight for organic carbon in the soil.

$$1 \text{ ml } N\ K_2Cr_2O_7 \equiv \frac{12}{4000} \times \frac{100}{77} \text{ or } 0{\cdot}0039 \text{ g carbon}$$

$$\%\text{ organic C} = \frac{\text{ml } K_2Cr_2O_7 \text{ reduced} \times N \times 0{\cdot}0039 \times 100}{\text{weight of soil}}$$

$$= \frac{(B - S) \times 0{\cdot}5 \times 0{\cdot}0039 \times 100}{\text{weight of soil}}$$

B and *S* are blank and sample titrations (ml) and *N*, the normality of Fe^{2+} solution. Organic matter is approximately %C x 1·724.

7.13 DETERMINATIONS OF ORGANIC AND NITRATE NITROGEN

Aside from fertiliser and minor mineral sources, soil nitrogen comes from plant proteins and their associated amino acids as well as from lignified tissue. Such organic nitrogen has all at one time been fixed from atmospheric nitrogen. Figure 7.7 shows the biogeochemical pathways of nitrogen. Nitrate is the most important form of nitrogen taken up by plants and is the form most rapidly leached from soils. Only a small amount of cyclic nitrogen may at any time be in this form but leaching of excess fertiliser may lead to wastage which can be monitored from runoff.

Although total nitrogen cannot be used alone as an index either of soil fertility or of nitrogen requirement, it does indicate the amount of nitrogen stored in the soil and potentially, if not

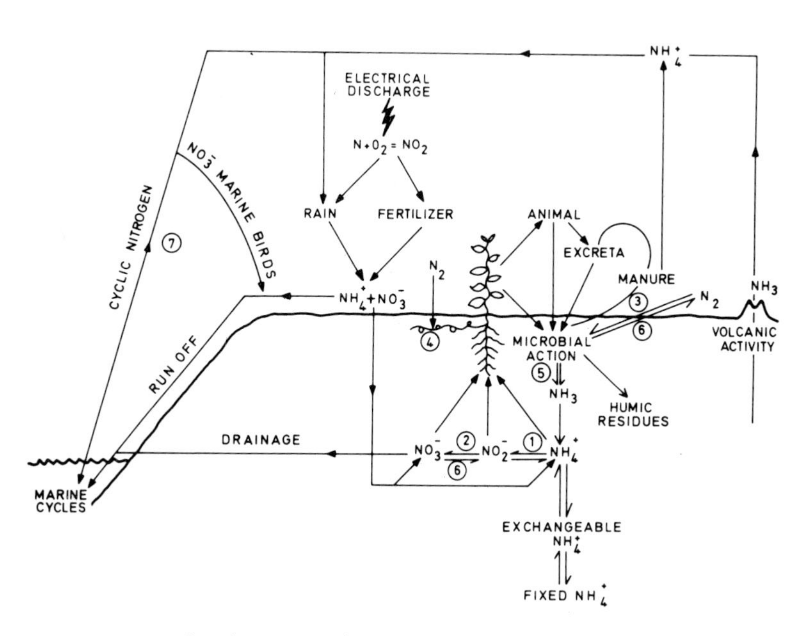

FIGURE 7.7 *The nitrogen cycle.*

1. Nitrosomonas, Nitrosococcus *(aerobic)* } *NITRIFICATION*
2. Nitrobacter *(aerobic)*
3. Azotobacter *(aerobic)*, Clostridium *(anaerobic)* } *NITROGEN FIXATION*
4. Rhizobium *(symbiotic)*
5. *many bacteria and other organisms* *AMMONIFICATION*
6. Pseudomonas, Micrococcus, Thiobacillus *(all anaerobic)* *DENTRIFICATION*
7. *N salts in rain water, sea spray etc.*

actually available. Available nitrogen is a very elusive quantity to pursue because of nitrogen transformations by bacteria with which plants themselves compete for nitrogen. Nitrogen starvation in a soil is normally associated with a high C:N ratio. Total nitrogen in temperate regions has often been used as an indicator of the humus content of soils, and under these conditions C:N ratios are commonly around 12:1 with nitrogen comprising about 5% of the humic matter.

Organic nitrogen The procedure described below is known as the Kjeldahl method, which recovers organic and ammoniacal forms of nitrogen. Nitrate is lost by volatilisation of nitric acid during treatment with sulphuric acid. The latter digestion converts organic nitrogen to an ammoniacal form. Ammonia gas is then determined by steam distillation followed by titration. For convenience, when dealing with large numbers of samples the required apparatus includes a Kjeldahl digestion rack and a micro-Kjeldahl apparatus consisting of a Markham distillation unit connected to a Liebig condenser and provided with steam generator and suction pump, as illustrated in Figure 7.8. This standard apparatus contributes to the reproducibility of results.

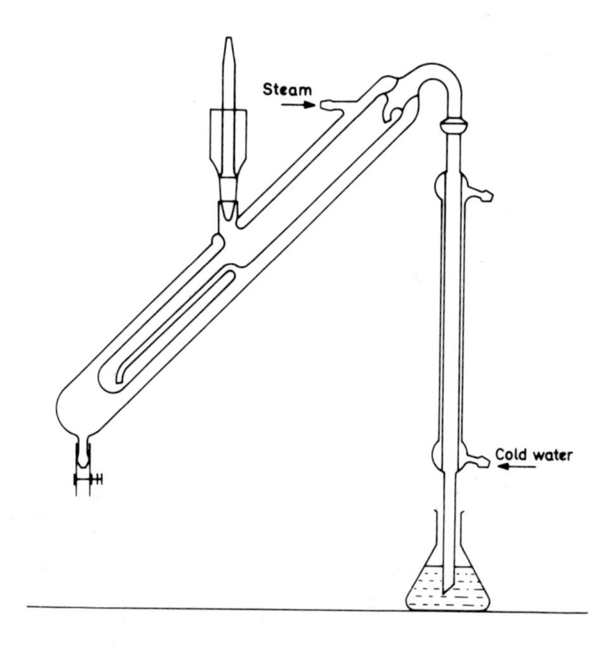

FIGURE 7.8 *The Markham distillation unit*

20 g air-dry fine earth (generally less for clays and organic soils) is placed in a 500 ml Kjeldahl flask. About 50 ml distilled water is added, followed by Kjedahl catalyst tablets. Tablets contain anhydrous sodium sulphate and smaller quantities of copper sulphate or selenium. Their total weight need not exceed 20 g. If tablets are not available then about 17 g anhydrous sodium sulphate and 3 g $CuSO_4$ (one or two crystals) can be added. The catalyst is a digestion accelerator, which raises the boiling point of concentrated sulphuric acid from 330°C into the range 360-420°C at which the digestion is most efficient.

35 ml nitrogen-free conc. H_2SO_4 is cautiously added while swirling the flask. Calcareous soils may give a vigorous reaction. The mixture is then heated on a Kjeldahl rack until the liquid turns a green or straw colour. Great care must be exercised with the heat control to ensure no loss of material due to frothing during this digestion. It is best to begin with a low heat, especially when working with a batch of flasks. After development of the straw colour the material is allowed to remain digesting for a further 15-20 minutes. The flask is then cooled, diluted with distilled water and its contents transferred to a 250 ml graduated flask. The Kjeldahl flask is rinsed several times to prevent loss of ammonium ion and the washings are added to the graduated flask which is then made up to 250 ml with distilled water.

For the estimation of organic nitrogen in runoff water, a 250 ml portion of the suspension (100 ml if over 5 g solids per 100 ml) is transferred to a Kjeldahl flask. Catalyst is added and then the liquid substantially evaporated before carefully adding the sulphuric acid.

10 ml of the digest are admitted into the boiling chamber of the steam distillation apparatus followed by 10 ml 40% NaOH. The receiving flask below the condenser should contain 10 ml 4% boric acid solution which includes solutions of methyl red (0·1%) and bromcresol green (0·5%) dissolved in 95% ethanol. 5 ml of this mixed indicator solution are added to each litre of boric acid solution so that about 3 drops will be adequate. The delivery tube from the condenser must dip into the liquid of the receiving flask and on distillation the boric-indicator solution will turn bluish. Distillation, which is effected in about 5 minutes, is begun by admitting steam into the boiling chamber. At the end of this time the receiving flask is removed for titration to be carried out, and the boiling chamber is pre-

pared for a repeat determination.

0·01*N* HC*l* is placed in a burette and titrated with the contents of the receiving flask. The end point is when the blue colour is first replaced by a pinkish red (about pH 4·5). The volume of HC*l* is noted and distillation of further aliquots of digest is carried out until two burette readings differ by less than 0·2 ml. A further distillation is then run using 10 ml distilled water instead of sample digest, and a blank titration carried out.

Volume of HC*l*(*V*) required for equivalence with ammonium ions = ml HC*l* in sample titration - ml HC*l* in blank titration.

1 litre *N* HC*l* ≡ 14 g nitrogen

1 ml 0·01*N* HC*l* $\equiv \frac{14}{1000 \times 100}$ or 0·14 mg nitrogen

Wt. of nitrogen in 250 ml digest from which a 10 ml sample was distilled $= V \times \frac{0 \cdot 14}{1000} \times \frac{250}{10}$ g

$$\% \text{ N in soil} = \frac{V}{\text{wt. of soil}} \times \frac{0 \cdot 14}{1000} \times \frac{250}{10} \times 100$$

$$= \frac{V \times 0 \cdot 35}{\text{wt. of soil}}$$

For a runoff sample, provided that 250 ml was taken originally and treated according to the procedure above, the concentration of N in the original runoff suspension

$$= V \times 0 \cdot 14 \times \frac{250}{10} \times 4 \text{ mg/litre}$$

A rough guide to the percentage organic matter in humid region soils is provided by %N × 20, while kg per hectare N in soil approximates to %N × 22,000.

Nitrate nitrogen This method is based on development of the nitrophenoldisulphonic acid-yellow colour, the strength of which is measured by light absorbance at 420 nm (Jackson, 1958).

Required reagents include phenol 2, 4-disulphonic acid; 25 g A.R. phenol is dissolved in 150 ml conc. H_2SO_4 and then 75 ml fuming H_2SO_4 is added. This mixture is heated in a boiling water bath for 2 hours. The resulting material is stored in a dark glass bottle and is extremely corrosive. Standard nitrate solution: 0·7221 g A.R. dry KNO_3 is dissolved in water and made up to 1 litre. This is equivalent to 100 p.p.m. This stock standard is then diluted 20 ml to 200 ml in a volumetric flask, giving 10 p.p.m. 2, 5, 10 and 15 ml aliquots of this solution are then placed in 8 cm evaporating dishes and reduced to dryness. Nitrate extraction solution: 200 ml *N* $CuSO_4$ solution (159·606 g $CuSO_4$ dissolved in 1 litre H_2O) is mixed with 1 litre 0·6% Ag_2SO_4 solution (6 g Ag_2SO_4 in 1 litre H_2O). The entire mixture is then diluted to 10 litres with water. The Ag_2SO_4 removes chlorides up to 338 p.p.m. or 0·03%. If the soil being tested has a high chloride value (e.g. saline soils) then 2·25 g powdered Ag_2SO_4 is initially mixed with the soil for every 1% chloride present.

For extraction of nitrate-nitrogen from soil samples, 50 g freshly dug soil with no material coarser than about 0·5 cm (25 g in the case of organic soil) is placed in a shaking bottle and a 250 ml portion of extraction solution added. A further weighed field sample is then oven-dried for moisture determination. After the test sample has been shaken mechanically for about 10 minutes, 0·4 g $Ca(OH)_2$ is added. Shaking is continued for a further 5 minutes and 1 g $MgCO_3$ is added, which clears the suspension by precipitating Cu and Ag as salts. The suspension is filtered and after rejecting the first runnings a 10 ml aliquot is delivered into an 8 cm evaporating dish.

For a runoff sample, 250 ml is made alkaline by addition of 0.25 g $CaCO_3$ powder and is then evaporated to dryness soon after collection. 100 ml extraction solution is then added and the material transferred to a shaking bottle for 5 minutes. The same procedure is then followed as for a soil sample except that about half quantities of $Ca(OH)_2$ and $MgCO_3$ are required. Also a 20 ml aliquot is evaporated and the final coloured solution is diluted to 40 ml.

To the evaporating dishes, 3 ml phenol 2, 4-disulphonic acid is added. The liquid is carefully swirled and 15 ml distilled water added. The solutions are thoroughly stirred and transferred to 100 ml graduated flasks. When cool, 6*N* NH_4OH is added slowly until a yellow colour has developed, and then a further 3 ml. The solutions are finally made to 100 ml with water (40 ml for runoff). The series of standards represent 0·2, 0·5, 1.0 and 1·5 p.p.m. nitrate-nitrogen.

A calibration curve is constructed with concentration (horizontal axis) plotted against optical density. Semi-logarithmic paper is useful for this purpose, with the log scale being used for optical density. Solutions are measured at 420 nm on a spectrophotometer or, alternatively, a colorimeter with a violet filter.

For soil samples, p.p.m. nitrate-nitrogen in oven-dry soil

= p.p.m. from curve × dilution of aliquot × soil dilution factor

$$= \text{p.p.m. from curve} \times \frac{\text{ml final yellow soln.}}{\text{ml aliquot evaporated}} \times \frac{\text{ml extracting soln.}}{\text{g oven-dry soil extracted}}$$

$$= \text{p.p.m. from curve} \times \frac{100}{10} \times \frac{250 + \text{moisture}}{50 - \text{moisture}}$$

For runoff, p.p.m. nitrate-nitrogen in original runoff suspension

$$= \text{p.p.m. from curve} \times \frac{\text{ml final yellow soln.}}{\text{ml aliquot evaporated}} \times \frac{\text{ml extracting soln.}}{\text{ml runoff evaporated}}$$

$$= \text{p.p.m. from curve} \times \frac{40}{20} \times \frac{100}{250} \quad \textit{or} \quad \text{p.p.m.} \times 0.8$$

There are now both more sophisticated and quicker methods for nitrate measurement. More accessible in cost is the use of an ion selective electrode (*see* Section 7.10, p.217).

7.14 HUMUS CHARACTERISATION

Although the term 'humus' has a specific meaning in relation to soil organic matter it nevertheless exists in various forms, free or combined, and there are gradations of material from senescent parts of plants to amorphous matter. Estimations of humus content are at best approximate, owing to the heterogeneous nature of the material. Traditionally, humus has been separated on the basis of alkali, acid and alcohol-soluble fractions (Figure 7.9), illustrating the different response of the various functional groups on humus molecules. As in other investigations, soil type determines which extractant will be most suitable. Sodium hydroxide may be used for acid soils with free humus of low molecular weight and low degree of condensation. For soils above about 5·5 pH an initial hydrochloric acid treatment will be necessary, while for soils containing free calcium, a mixture of sodium hydroxide and sodium pyrophosphate will be found most efficient (reagent details with method C).

The four methods which follow include two for the estimation of humus as a percentage of the soil material, method A being the standard gravimetric method suitable for all soils. Humus extracts, providing they are dilute and optically homogeneous, can be studied using a colorimeter or spectrophotometer. Since significant variation in light absorbance is noted between humus of low and high molecular weight, the simpler forms of this method (described under C and D) can be used to assess humus *type* or *quality* over a range of soils. This index provides a further criterion on which to base soil classification. Method B is a quantitative adaptation of the colorimetric method, though it is only suitable for acid soils because variations in humus type complicate the interpretation. The colorimetric method offers advantages with larger numbers of samples.

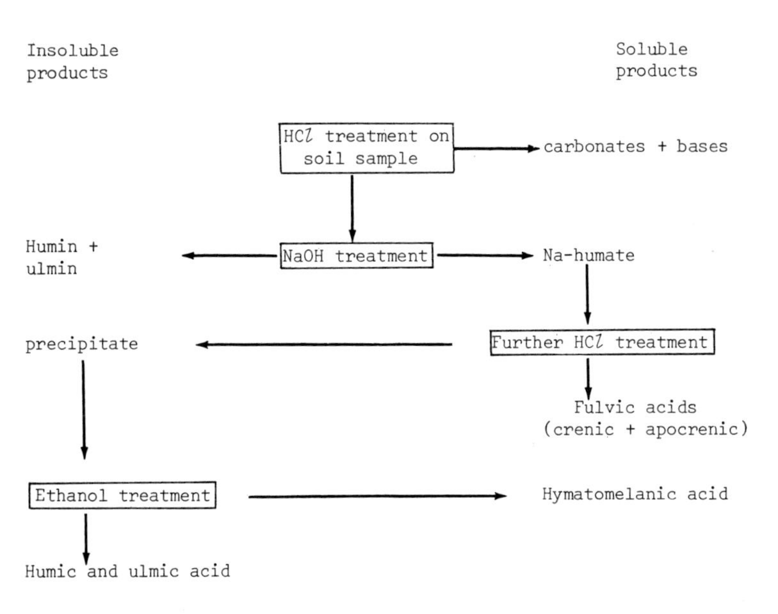

FIGURE 7.9 *Fractionation of humus*

Method A 50 g of oven-dry fine earth is placed in a 250 ml beaker. 100 ml 5% HCl is added (only necessary if pH above about 5·5), the mixture warmed and then cooled and filtered (or centrifuged). The soil is then leached (or washed) until acid free (tested with litmus or universal indicator). Soil collected on the filter (or compacted residue from centrifugation) is then returned to the 250 ml beaker. 100 ml 5% NH_4OH is added and the mixture boiled for about 5 minutes. The suspension is filtered (or centrifuged) and the solid residue in either case washed to avoid loss of ammonium humate. The humate solution is finally treated with 5% HCl until just acid (tested with litmus or universal indicator), at which point a precipitate of humic substances will form. The material is then filtered through a weighed filter paper. The filter paper is dried in an oven and weighed to obtain the weight of humic matter.

$$\% \text{ humus} = \frac{\text{wt of humic residue} \times 100}{\text{wt of sample}}$$

Note: filtration is often time-consuming and while centrifugation using 50-250 ml tubes is the obvious answer, circumstances may dictate either double filtration through coarse and fine papers or suction filtration using a Buchner funnel and side-arm flask.

Method B Two 5 g samples of peat are boiled for 15 minutes with 30 ml 10% NaOH. The extracts are then filtered (or centrifuged) and one of them reduced to dryness followed by ignition so that the weight of humus can be determined. When this is known, the concentration of the second extract is adjusted so that it contains exactly 1% humic matter. This becomes the stock standard from which dilutions are made to cover the range 0·01 to 0·001% humus. A calibration curve of humus percentage against colorimeter readings (absorbance) is constructed using fresh solutions and zero set with NaOH. A violet filter is used (Cornwall, 1958).

1 g of air-dry fine earth sample is placed in a small conical flask or test tube and boiled for a few minutes with about 20 ml 10% NaOH. 10 ml of the filtrate (or supernatant) from this extraction is diluted until its absorbance value fits the range of the calibration curve.

Humus % = value from curve × soil:extractant ratio × further dilution

Method C This is based on the discovery that the ratio of optical density (otherwise known as absorbance or extinction) at 465 nm to that at 665 nm (the $E_4:E_6$ ratio, see Table 7.7) is independent of carbon content while remaining characteristic of different types of humus (Kononova, 1966). Approximately 5 g of the material to be investigated is placed in a shaking bottle and shaken mechanically for about 15 minutes with 100 ml freshly prepared sodium hydroxide-sodium pyrophosphate solution (44·6 g $Na_4P_2O_7.10H_2O$ and 4 g NaOH dissolved in one litre of water). After allowing the extract to stand overnight it is then shaken briefly, transferred to 50 ml centrifuge tubes and spun at 3000 r.p.m. for 10 minutes. The optical density of the extract is then determined on the above wavebands using a spectrophotometer with 1 cm glass cells and

Soil type	$E_4:E_6$ ratio
Chernozems	3·0-3·5
Grey forest	3·5
Chestnut soils	3·8-4·0
Serozems	4·0-4·5
Podzolic soils	5·0
Fulvic acids	6·0-8·5

TABLE 7.7 *Absorbance ratios of selected soil types*

using reagent for setting zero on the instrument.

Clear differences will not be recognisable unless comparisons are made between different types of soils. For intergrading situations or for soils of very similar genesis the full method is required, involving the construction of spectrophotometric

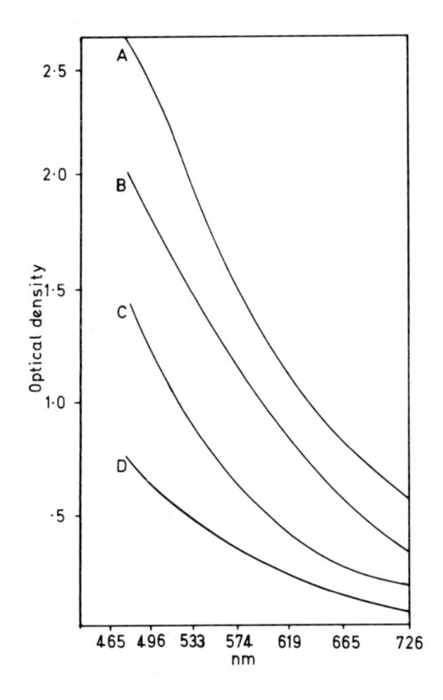

FIGURE 7.10 *Guide to the light absorption of humic extracts from selected soil types.* A, *chernozemic and brown calcareous soils;* B, *grey forest and most brown earths;* C, *red earths and brown podzolics;* D, *podzols and peat.*

curves (Figure 7.10) between 465 and 726 nm (this curve can also be constructed by using a series of coloured filters).

Method D This is based on the observation that the ratio of the amount of humic material dissolved by NaF to that dissolved by NaOH is characteristic of the type of humification in a soil. This method has been developed by Hargitai (1971) and is a modification of that devised by Hock (1936).

Two 2 g air-dry samples of the soil to be investigated are placed in separate shaking bottles which are labelled NaOH and NaF. To one, 20 ml 0·5% NaOH is added and to the other, 20 ml 1% NaF. The two bottles are sealed, shaken mechanically and centrifuged as in method C. Alternatively, after allowing to stand covered for 48 hours the extracts are passed through a fine filter paper. Optical densities are then measured at 436 nm, 465 nm, 496 nm, 533 nm, 574 nm, 619 nm, 666 nm, 726 nm and 750 nm using a spectrophotometer with 1 cm glass cells. If extracts are too strong, they must be diluted and the dilution accounted for in the following calculation:

$$\text{Humus stability index } (Q) = \frac{\text{optical density NaF extract}}{\text{optical density NaOH extract}}$$

$$\text{e.g.}\quad Q_{436} = \frac{1\cdot500}{3\cdot000} = 0\cdot500$$

The characterising value for a soil is the average index for the nine wavelengths measured,

$$\text{thus } \overline{Q} = \frac{\sum Q_{436} - Q_{750}}{9}$$

This method can be adapted for a field test. Although subjective the test can provide a quick indication of humus quality as follows:

NaOH extract darkest	$Q < 1$	weak
Two extracts identical	$Q = 1$	medium
NaF extract darkest	$Q > 1$	good

A further value, known as the stability coefficient (K), is obtained by dividing $\overline{Q}$ by the organic matter content of the soil. The latter is expressed as a percentage of the air-dry soil and for this determination Hargitai has used the chromic acid oxidation method of Tyurin (1931). The method described in Section 7.12 would be equally appropriate. K is perhaps a better guide to the quality of humus in relation to soil genesis, biochemistry and classification (Table 7.8).

Value of K	Type of soil/material
< 0·001	Lignite and coals
0·001	Mor humus, forest litter
0·01	Peats, manure, some alkali soils
0·1	Moor, meadow, podzols, podzolised brown earths
1·0	Brown forest soils
10·0	Chernozems

TABLE 7.8 *Humus coefficients of selected soils*

7.15 EXTRACTION AND IDENTIFICATION OF POLLEN GRAINS AND SPORES

Samples for pollen analysis are collected as described in Chapter 2. The following schedule assumes that a given sample contains a mixture of materials needing to be removed before microscopical study is possible. In practice samples usually contain one main constituent so that one or more stages can be eliminated. As the method makes considerable use of the centrifuge, it is sensible to treat samples in batches of up to eight. The centrifuge should take 20 ml and 50 ml tubes and desirably 100 ml tubes or 250 ml buckets as well. The size of vessels depends largely on the material being processed, and after centrifuging, the supernatant is discarded in every case (Smith, 1966).

Humified organic matter Between 0·5 and 10 g of sample (larger quantity for mineral material) is placed in conical flasks. About 30 ml 10% NaOH is added and the flasks boiled gently for 10-15 minutes on a hotplate preferably in a fume cupboard. After cooling, the contents are decanted through 72 mesh nylon sifting cloth (200 μm apertures) into beakers. The filtrates are swirled and passed through 122 mesh cloth (115 μm apertures) into 100 ml centrifuge tubes. This removes all particles larger than fine sand and allows silt, clay, pollen and other microscopical plant and animal remains to pass through. The filtrates are diluted to 100 ml with distilled water, then centrifuged for one minute at 2000 r.p.m. Peaty samples often need no further treatment.

Calcareous material Samples are treated with 10% HCl. Providing very little material remains, the final treatment can then follow.

Siliceous material Samples should be dealt with either by heavy liquid separation or, as here, by treatment with hydrofluoric acid. Residues from NaOH treatment are transferred to 30 ml nickel crucibles, if possible, without water. 10-15 ml 40%

HF is added cautiously. The crucibles are placed on a hotplate in a fume cupboard and allowed to boil for 5-10 minutes. While still in the fume cupboard the crucible contents are transferred to 50 ml polypropylene tubes and diluted with warm 10% HCl. They are then stirred, centrifuged, washed with warm water and centrifuged again. For strictly quantitative studies the whole operation can be conveniently carried out using polypropylene tubes and heating in an oil bath.

Cellular material Residues from earlier treatments must be transferred to glass tubes, dehydrated with 5-10 ml glacial acetic acid in a fume cupboard, then centrifuged. 5-10 ml freshly prepared acetylation reagent is then added as follows: acetic anhydride (9 parts) followed by 98% H_2SO_4 delivered from a burette (1 part). The mixtures are boiled in an oil bath for a few minutes until the brownish colour of cellulose triacetate is observed. After adding glacial acetic acid the tubes are centrifuged, washed again with glacial acetic acid and centrifuged. Residues are now washed with water and centrifuged again (Erdtman,1960; Kummel and Raup, 1965).

Final treatment All residues must be made alkaline before they will absorb stain. Washed residues are therefore heated for about 10 minutes with 10% NaOH, centrifuged, washed with water and centrifuged finally. The residue is then taken up with a few ml 50:50 glycerin and water mixture or glycerol jelly. Each 100 ml of this mounting medium should contain 2 ml, 1% water soluble safranin dye. Dilution of the residue at this stage is dependent upon the concentration of pollen and its clarity within the extract.

Identification Figure 7.11 shows a selection of pollen and spore types. Although the number of pollen types in a vertical profile can be considerable it is often the case that in any one sample the number does not exceed about 30 and is often less. Identification is on the basis of size, shape, numbers of pores and/or furrows and on the structure and sculpturing of the outer wall or exine. Guides to pollen identification are provided in Erdtman *et al.* (1961, 1963) and Faegri and Iversen (1964), while broader treatments of pollen analysis and vegetation history, incorporating discussion of macrofossils, include Godwin (1956), West (1968), Pennington (1969), Dimbleby (1970), Walker and West (1970) and Moore and Bellamy (1973).

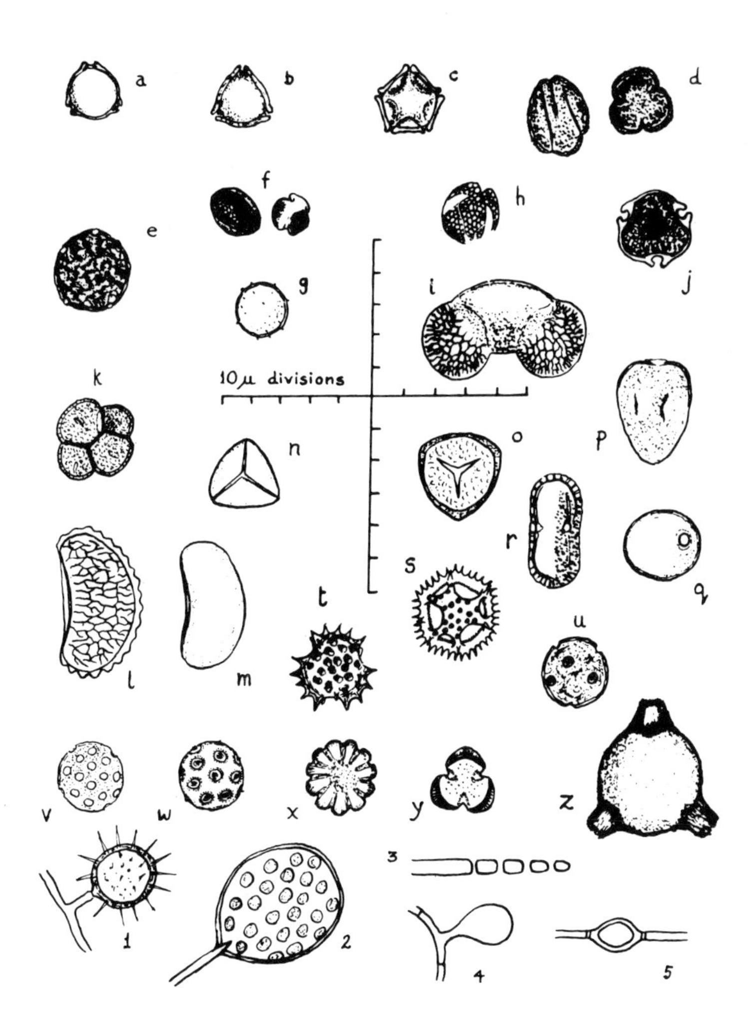

FIGURE 7.11 *Some examples of pollen grains and spores*
a, Betula pubescens *(birch tree)*; b, Corylus avellana *(hazel)*; c, Alnus glutinosa *(alder)*; d, Quercus *(oak)*; e, Ulmus *(elm)*; f, Salix *(willow)*; g, Juniperus *(juniper)*; h, Fraxinus *(ash)*; i, Pinus *(pine)*; j, Tilia *(lime)*; k, Ericaceae *(heaths)*; l, Polypodium *(polypody fern)*; m, Dryopteris-type *(male fern)*; n, Pteridium aquilinum *(bracken)*; o, Sphagnum *(bog moss)*; p, Cyperaceae *(sedges)*; q, Gramineae *(grasses)*; r, Umbelliferae:Heracleum-type *(hog weed)*; s, *Compositae-Liguliflorae:* Taraxacum-*type (dandelion)*; t, *Compositae-Tubuliflorae:* Bellis-*type (daisy)*; u, Plantago lanceolata *(ribwort plantain)*; v, Chenopodium album *(fat hen)*; w, *Caryophyllaceae e.g.* Lychnis flos-cuculi *(ragged robin)*; x, Polygala vulgaris *(milkwort)*; y, Artemisia *(wormwood)*; z, Chamaenerion angustifolium *(rosebay willowherb)*; *Examples of sporogenesis in the fungi: 1, aleuriospore; 2, sporangium with sporangiospores; 3, arthrospore; 4, conidiospore; 5, chlamydospore.*

Counting Pollen slides are scanned in a series of traverses until a statistically valid number of grains has been counted. 200-500 grains can be regarded as an adequate number for most purposes. While initial observations may be carried out at lower power, counting is normally carried out at about 400 times magnification (eyepieces × 10 and objective × 40), with higher magnification and oil immersion required in problem cases. The data for each pollen type is recorded as a percentage of the pollen sum or as a percentage of the total tree pollen. While the former is of greater use in picturing ecological change the latter has been widely used for purposes of post-glacial pollen zonation. The data is presented as separate graphs or histograms with percentage plotted against depth. Such composite illustrations are known as pollen diagrams.

Interpretation Owing to variables of pollen production, dispersal and even preservation, there are difficulties in regarding pollen data as an entirely accurate reflection of former vegetation. On many bogs, pollen from on-site vegetation swamps that blowing in from adjacent dry land habitats. It is thus difficult to estimate the size of area which is reliably represented by the pollen in a particular core, though the seriousness of this problem will depend on whether the site itself or the wider locality is the subject of the investigation. In this connection, studies of pollen transfer are proving to be of value in the interpretation of fossil pollen records (Tauber, 1967; Tinsley and Smith, 1974).

REFERENCES

Ball, D.F. 'Loss on ignition as an estimate of organic matter and organic carbon in non-calcareous soils', *J. of Soil Sci.* **15** (1964), 84-92.

Bascomb, C.L. 'A calcimeter for routine use on soil samples', *Chem. & Ind.*, **45** (1961), 1826-7.

Bear, F.E. *Chemistry of the soil*, (2nd Edn.) Van Nostrand Reinhold, New York, 1964.

Black, C.A., Evans, D.D., Ensminger, L.E., Clark, F.E. and White, J.L. *Methods of soil analysis* (Vol. 2) American Society of Agronomy, Madison, 1965.

Bower, C.A. and Huss, R.B. 'Rapid conductometric method for estimating gypsum in soils', *Soil Sci.*, **66** (1948), 199-202.

Boyd, D.A. *Soil phosphorus*, Ministry of Agriculture, Technical Bulletin No. 13, H.M.S.O., 1962.

British Standards Institution, *Methods of testing soils for civil engineering purposes*, London, B.S. 1377, 1967.

Coffin, D.E. 'A method for the determination of free iron in soils and clays', *Can. J. Soil Sci.*, **43** (1963), 7-17.

Cornwall, I.W. *Soils for the archaeologist*, Phoenix, London, 1958.

Davies, B.E. 'Anomalous levels of trace elements in Welsh soils', *Welsh Soils Discussion Group Report No. 9*, (1968), 72-88.

Dimbleby, G.W. *Plants and archaeology*, John Baker, London, 1970.

Erdtman, O.G.E. 'The acetolysis method', *Svensk. Bot. Tidskr*, **54** (1960), 561.

Erdtman, O.G.E., Berglund, B. and Praglowski, J. *An introduction to a Scandinavian pollen flora I*, Almqvist and Wiksells, Stockholm, 1961.

Erdtman, O.G.E., Praglowski, J. and Nilsson, S. *An introduction to a Scandinavian pollen flora II*, Almqvist and Wiksells, Stockholm, 1963.

Faegri, K. and Iversen, J. *Textbook of pollen analysis*, Blackwell, Oxford, 1964.

Gilchrist-Shirlaw, D.W. *A practical course in agricultural chemistry*, Pergamon, Oxford, 1967.

Godwin, H. *History of the British flora*, Cambridge University Press, London, 1956.

Hargitai, L. 'The effect of soil genetical and biochemical factors on humus quality', *Transactions of the Prague symposium "Humus et Planta V"*, 1971.

Hesse, P.R. *A text book of soil chemical analysis*, John Murray, London, 1971.

Hock, A. *Z. PflErnähr. Düng*, **47** (1936), 304.

Jackson, M.L. *Soil chemical analysis*, Constable, London, 1958.

Kummel, B. and Raup, D. *Handbook of paleontological techniques*, Freeman, San Francisco, 1965.

Kononova, M.M. *Soil organic matter*, Pergamon, Oxford, 1966.

Larsen, S. 'Soil phosphorus' *Adv. in Agronomy* 19; 151-210 1967

MacPhee, W.S.G. and Ball, D.F. 'Routine determination of calcium and magnesium in soil extracts by atomic absorption spectrophotometry', *Journal of the Science of Food and Agriculture*, 18 (1967), 376-80.

Metson, A.J. *Methods of testing soils for soil survey samples*, New Zealand D.S.I.R. Soils Bulletin No. 12, 1961.

Ministry of Agriculture, Fisheries and Food, *Fertilizer recommendations*, Bulletin 209, H.M.S.O., 1967.

Moore, P.D. and Bellamy, D. *Peatlands*, Elek, London, 1973.

Pennington, W. *The history of British vegetation*, E.U.P., London, 1969.

Piper, C.S. *Soil and plant analysis*, University of Adelaide, Adelaide, 1942.

Purves, D. Contamination of urban garden soils, *Nature*, 210 (1966), 1077 and see also *J. of soil sci.*, 20 (1969), 288.

Smith, R.T. 'Some refinements in the technique of pollen and spore extraction from soil', *Laboratory Practice*, 15 (1966), 1120-23.

Tauber, H. 'Differential pollen dispersion and filtration' in *Quaternary Palaeoecology* Vol. 7 Eds. Cushing, E.J. and Wright, H.E. Jr. Yale, 1967.

Tinsley, H.M. and Smith, R.T. 'Surface pollen studies across a woodland/heath transition and their application to the interpretation of pollen diagrams', *New Phytol.*, 73 (1974), 549-67.

Tintometer Ltd. Salisbury, England, *Colorimetric chemical analytical methods*.

Tyurin, I.V. 'A modification of a volumetric method of humus determination with chromic acid', *Pochvovedenie*, 5/6 (1931), 36.

U.S.D.A., Soil Survey Staff: *Soil Survey Manual*, Agricultural Handbook No. 18, Washington D.C., 1951.

U.S.D.A. Handbook No. 60, *Diagnosis and improvement of saline and alkali soils*, Ed. L.A. Richards, 1954.

Walker, D. and West, R.G. Eds. *Studies in the vegetational history of the British Isles*, C.U.P., London, 1970.

Walkley, A.J. and Black, I.A. 'An examination of the Degtjareff method for determining soil organic matter and a proposed modification of the chromic acid titration method', *Soil Sci.*, 37 (1934), 29-38.

Walsh, L.M., Ed. *Instrumental methods for analysis of soils and plant tissue*, Soil Science Society of America, Madison, 1971.

Warren, H.V. 'Geology, trace elements and epidemiology', *Geographical Journal*, 130 (1964), 525-28.

West, R.G. *Pleistocene geology and biology*, Longmans, London, 1968.

A SELECTED ADDRESSES

Australia

C.S.I.R.O. Soil Publications, Melbourne.

Belgium

Soil Survey Centre, Université de Ghent, Rozier 6, Ghent.

Canada

Canada Department of Agriculture, Ottawa.

England and Wales

Rothamsted Experimental Station, Harpenden, Herts.

France

Institut National de Recherche Agronomique,
Etoile de Choisy, Route de saint Cyr, Versailles, Seine et Oise.

Germany

Institut für Bodenkunde, Stuttgart-Hohenheim.

Hungary

Institute for Soil Science and Agricultural Chemistry, Budapest.

Ireland

An Foras Taluntais, 33 Merrion Road, Dublin 4.

Netherlands

State Agricultural University (Landbouhoogeschool),
Gen. Foulkesweg la, Wageningen,
also Netherlands Soil Survey Institute, Wageningen.

New Zealand

New Zealand Department of Scientific and Industrial Research, Soil Bureau, Lower Hutt.

Portugal

Serviço de Reconhecimento e Ordenamento Agrério, Av. Duque d'Avila, 32-2°, Lisbon-1.

Rumania

Geological Institute, Bucharest.

Scotland

Macaulay Institute for Soil Research, Craigiebuckler, Aberdeen.

Spain

Instituto de Edafologia, C.S.I.C., Madrid.

U.S.A.

U.S.D.A. Federal Center Building, Hyattsville, Md. 20782.

U.S.D.A. Soil Conservation Service, Washington D.C. 20250.

U.S.S.R.

Dokuchaiev Soil Institute, Pygevski per. 7, Moscow.

International organisations

International Society of Soil Science *and* Royal Tropical Institute, 63, Mauritskade, Amsterdam, Netherlands.

Office de la Recherche Scientifique et Technique Outre-Mer, (O.R.S.T.O.M.) 70 route d'Aulnay, Bondy, Seine, France.

World Soil Resources Office, Soil Resources and Survey Branch, F.A.O., Via delle Terme di Caracalla, 00100, Rome.

For addresses of all 'National Societies of Soil Science' see Bulletin of the I.S.S.S. No. 45, 1974.

B UNITS AND ABBRIEVIATIONS

This text spans a wide range of activities and embodies the units of both the British and metric systems with which workers in particular fields are most familiar. As this is primarily an introductory text and as progress in the adoption of SI units has not been as rapid as expected the authors have felt it would be unsuitable to maintain rigid consistency in the use of units.

A.R. Analytical grade reagent

C.E.C. Cation exchange capacity

conc. Concentrated reagent

e.s.d. Equivalent spherical diameter (particles)

E.W. Equivalent weight

S.G. Specific gravity (density relative to water)

soln. Solution

V:V Volume to volume for dilution

W:V Weight to volume for solution and dilution

m.e. (m.eq.) Milliequivalent - equivalent weight in mg

m.e./100 g Milliequivalents per 100 grams (ion exchange)

mg Milligrammes, 10^{-3} g

μg Microgrammes, 10^{-6} g

ml Millilitre, cc (or cm^3 in SI units)

g/litre Grams per litre (1000 ml)(or g dm^{-3} in SI units)

p.p.m. Parts per million e.g. mg/litre (or μg cm^{-3} in SI units)

μ; μm Micron or micrometre; 0·001 mm (sieve meshes and microscopy)

nm	Nanometre, 10^{-9} metre, mμ, 10 Angstroms(colorimetry)
mho	1/ohm, (or 1 Siemens, 1 S, in SI units)
mmhos/cm	Millimhos per cm (mS cm^{-1})(conductivity)
E.c.	Electrical conductivity
r.p.m.	Revolutions per minute (centrifugation)
S.P.	Spectrographically pure reagent

C STRENGTHS OF SOLUTIONS

The concentration of solutions can be expressed in a number of ways including percent by weight, parts per million or specific gravity, but it has also been standard practice in many determinations to express concentration in terms of chemical equivalence or normality. A normal (*N*) solution is one which contains the equivalent weight of substance (solute) in grams, per litre of solvent (usually water). A molar (*M*) solution is one which contains the molecular weight of a substance (Appendix F) in g/litre and therefore molar and normal solutions (or fractions thereof) will often be identical. Although the present trend in analytical chemistry is towards standardisation in terms of molarity, most sources still refer to normality which is why we have continued its use here.

It should be realised, however, that in many procedures reference to normality is merely a guide to the concentration level needed for efficiency of reaction and for this purpose a weight: volume percentage could equally well be used. For titrimetric determinations, however, standard solutions of given normality are essential.

The following table is intended to assist in preparing some of the more common reagents.

Reagent	Approx. normality of conc. reagent	% by wt.	S.G.	ml. conc. reagent/ litre for N soln.	g reagent in 1 litre N soln.
Acetic acid, CH_3COOH	17·2	99·5	1·05	58	60·0
Ammonium hydroxide, NH_4OH	14	27	0·88	71	35·0
Hydrochloric acid, HCl	11·2	36	1·18	89	36·5
Hydrofluoric acid, HF	26	46	1·15	38	20·0
Nitric acid, HNO_3	16	70	1·42	63	63·0
Perchloric acid, $HClO_4$	11·6	72	1·66	86	100·5
Phosphoric acid, H_3PO_4	45	85	1·71	22	32·7
Sulphuric acid, H_2SO_4	35·7	96	1·84	28	49·0
Ammonium acetate, CH_3COONH_4					77·1
Potassium hydroxide, KOH					56·0
Sodium hydroxide, $NaOH$					40·0

D pH INDICATORS

Indicator	Critical pH[a]	pH range	Colour change	Preparation
Malachite green[b]	1·0[c]	0·5-2·0	yellow/green/ blue green	A 2·68
Thymol blue[b]	1·9	1·2-2·8	red/orange/ yellow	A 2·15
Methyl orange	3·7[c]	3·1-4·4	red/orange	B
Bromphenol blue	4·0	3·0-4·6	yellow/blue/ violet	A 1·49
Bromcresol green	4·6	3·8-5·4	yellow/green/ blue	A 1·43
Chlorphenol red	5·6[c]	4·8-6·8	yellow/orange/ violet	A 2·38
Methyl red	5·7[c]	4·4-6·2	red/orange/ yellow	C
Bromcresol purple	6·2	5·2-6·8	yellow/purple/ violet	A 1·85
Bromthymol blue	6·9	6·0-7·6	yellow/green/ blue	A 1·60
Phenol red	7·3[c]	6·8-8·4	yellow/orange/ violet	A 2·82
Cresol red	7·8[c]	7·2-8·8	yellow/red	A 2·62
Phenolphthalein	8·3	8·2-10·2	colourless/ pink	D
Thymol blue[b]	8·9	8·0-9·6	yellow/purple/ violet	A 2·15
Thymolphthalein	9·4	9·2-11·5	colourless/ blue	D
Alizarin yellow R	10·3	10·0-11·5	yellow/orange/ red	A 3·48
Malachite green[b]	11·5[d]	10·5-12·0	blue green/ colourless	A 2·68

a most pronounced colour change
b weak dibasic acids with two ranges
c mid-point of gradual colour change
d permanently decolourised above this value
A prepared in aqueous solution (0·04%) after forming the Na-salt. 0·1 g dye is ground with the stated number of ml 0·1*N* NaOH (see right hand column) and is then made up to 250 ml with distilled water.
B 0·1 g dye is dissolved in 100 ml water
C 0·1 g is dissolved in 100 ml 60% ethanol - colour change is enhanced by addition of 0·8% methylene blue chloride
D 1.0 g dye is dissolved in 100 ml 90% ethonol

A number of the above dyes are available in water-soluble form which will simplify preparation.

'Full-range' and 'soil' pH indicators are available commercially but the following universal (mixed) indicator solution can be easily prepared in the laboratory. 60 mg methyl orange, 40 ml methyl red, 80 mg bromthymol blue, 100 mg thymol blue and 20 mg phenolphthalein are weighed out and dissolved in 100 ml 50% ethanol to which 0·1*N* NaOH is added dropwise until a yellow colour develops. The resulting indicator solution will provide the following approximate steps:

pH value	Colour
1-2	red
3-4	red/orange
5	orange
6	yellow
7	yellow/green
8	green
9	blue/green
10	blue

E TESTING OF ANALYTICAL RESULTS

Analytical methods depend for their success upon accuracy and precision. *Accuracy* describes the agreement between the experimentally determined value and the true or theoretical value. *Precision* is a measure of the ability of a method (or instrument) to give reproducible results irrespective of its accuracy. There may be many sources of error to account for such deviations about the true value and for varying degrees of reproducibility. These errors are described as *systematic* if they originate from personal errors of a non-accidental kind, from defects in instruments or reagents and through other routine sources such as assumptions which are made about the action of apparatus or reagents in a given procedure. *Random* errors, on the other hand, are often not able to be allocated to a particular cause, are often virtually unavoidable and are regarded as a matter of chance; but random errors are usually quite minor. Serious chance errors which are noticed will necessitate a repeat determination.

The *best estimate* or *mean* ($\bar{x}$) of a number of separate determinations is given by

$$\bar{x} = \frac{1}{n} \Sigma x_i$$

in which n is the number of values and Σx_i the sum of the individual values.

The reliability of a set of values i.e. their spread about the mean value, is measured by the *standard deviation*, S, in which

$$S = \sqrt{\left(\frac{\sum(x_i - \bar{x})^2}{n - 1}\right)}$$

while computation is easier in the following form:

$$S = \sqrt{\left(\frac{\Sigma x_i^2 - \left[(\Sigma x_i)^2/n\right]}{n - 1}\right)}$$

A useful way of comparing the spread of values obtained by two methods or for two different soils is given by the *coefficient of variation*, in which

$$CV = \left(\frac{S}{\bar{x}} \times 100\right)\%$$

The reliability of the mean, measured in terms of its *standard error*, $S_{\bar{x}}$, is given by the standard error of one determination, S_x, divided by the square root of the number of determinations, hence

$$S_{\bar{x}} = \frac{S_x}{n^{\frac{1}{2}}} \quad \text{or} \quad \sqrt{\left(\frac{\sum(x_i - \bar{x})^2}{n(n - 1)}\right)}$$

which is easier to compute in the form:

$$S_{\bar{x}} = \sqrt{\left(\frac{(\Sigma x_i)^2 - \left[(\Sigma x_i)^2/n\right]}{n(n - 1)}\right)}$$

A device enabling a statistical discrimination between two methods or two soils i.e. to tell whether the difference in the sets of results is *significant* and could not have arisen by accident (probabilities below 0·05 or 5 chances in 100) is given by the *Student's t test* which compares the standard errors of their respective means as follows:

$$t = \frac{\bar{x}_A - \bar{x}_B}{\sqrt{(S_{\bar{x}_A})^2 + (S_{\bar{x}_B})^2}}$$

where $S_{\bar{x}_A}$ is the standard error of one set of results (A), and $S_{\bar{x}_B}$ is the standard error of the other set (B).

F ATOMIC WEIGHTS OF THE ELEMENTS

Name	Symbol	Atomic number	Atomic weight
Actinium	Ac	89	227
Aluminum	Al	13	26.98
Americium	Am	95	243
Antimony	Sb	51	121.76
Argon	Ar	18	39.944
Arsenic	As	33	74.91
Astatine	At	85	[210]
Barium	Ba	56	137.36
Berkelium	Bk	97	[249]
Beryllium	Be	4	9.013
Bismuth	Bi	83	209.00
Boron	B	5	10.82
Bromine	Br	35	79.916
Cadmium	Cd	48	112.41
Calcium	Ca	20	40.08
Californium	Cf	98	[249]
Carbon	C	6	12.010
Cerium	Ce	58	140.13
Cesium	Cs	55	132.91
Chlorine	Cl	17	35.457
Chromium	Cr	24	52.01
Cobalt	Co	27	58.94
Copper	Cu	29	63.54
Curium	Cm	96	[245]

Name	Symbol	Atomic number	Atomic weight
Dysprosium	Dy	66	162.46
Einsteinium	Es	99	[254]
Erbium	Er	68	168.94
Europium	Eu	63	152.0
Fermium	Fm	100	[255]
Fluorine	F	9	19.00
Francium	Fr	87	[223]
Gadolinium	Gd	64	156.9
Gallium	Ga	31	69.72
Germanium	Ge	32	72.60
Gold	Au	79	197.0
Hafnium	Hf	72	178.6
Helium	He	2	4.003
Holmium	Ho	67	164.94
Hydrogen	H	1	1.0080
Indium	In	49	114.76
Iodine	I	53	126.91
Iridium	Ir	77	192.2
Iron	Fe	26	55.85
Krypton	Kr	36	83.80
Lanthanum	La	57	138.92
Lead	Pb	82	207.21
Lithium	Li	3	6.940
Lutetium	Lu	71	174.99
Magnesium	Mg	12	24.32
Manganese	Mn	25	54.94
Mendelevium	Md	101	[256]
Mercury	Hg	80	200.61
Molybdenum	Mo	42	95.95
Neodymium	Nd	60	144.27
Neon	Ne	10	20.183
Neptunium	Np	93	[237]
Nickel	Ni	28	58.69
Niobium	Nb	41	92.91
Nitrogen	N	7	14.008
Nobelium	No	102	[253]
Osmium	Os	76	190.2
Oxygen	O	8	16.000
Palladium	Pd	46	106.7

Name	Symbol	Atomic number	Atomic weight
Phosphorus	P	15	30.975
Platinum	Pt	78	195.23
Plutonium	Pu	94	[242]
Polonium	Po	84	210
Potassium	K	19	39.100
Praseodymium	Pr	59	140.92
Promethium	Pm	61	[145]
Protactinium	Pa	91	231
Radium	Ra	88	226.05
Radon	Rn	86	222
Rhenium	Re	75	186.31
Rhodium	Rh	45	102.91
Rubidium	Rb	37	85.48
Ruthenium	Ru	44	101.1
Samarium	Sm	62	150.43
Scandium	Sc	21	44.96
Selenium	Se	34	78.96
Silicon	Si	14	28.09
Silver	Ag	47	107.880
Sodium	Na	11	22.991
Strontium	Sr	38	87.63
Sulphur	S	16	32.066
Tantalum	Ta	73	180.95
Technetium	Tc	43	[99]
Tellurium	Te	52	127.61
Terbium	Tb	65	158.93
Thallium	Tl	81	204.39
Thorium	Th	90	232.05
Thulium	Tm	69	169.4
Tin	Sn	50	118.70
Titanium	Ti	22	47.90
Tungsten	W	74	183.92
Uranium	U	92	238.07
Vanadium	V	23	50.95
Xenon	Xe	54	131.3
Ytterbium	Yb	70	173.04
Yttrium	Y	39	88.92
Zinc	Zn	30	65.38
Zirconium	Zr	40	91.22

Notes

1. Atomic weights are based on the oxygen isotope 16.000.
2. Values in brackets indicate the mass number of the most stable isotope i.e. that which has the longest half-life.
3. Natural variation in the relative abundance of sulphur isotopes causes a range of ±0.003 in its atomic weight.
4. For most purposes atomic weights can be rounded up or down to the nearest whole number or half.

INDEX